# The Other of Climate Change

## CHALLENGING MIGRATION STUDIES

This provocative new series challenges the established field of migration studies to think beyond its policy-oriented frameworks and to engage with the complex and myriad forms in which the global migration regime is changing in the twenty-first century. It proposes to draw together studies that engage with the current transformation of the politics of migration, and the meaning of 'migrant,' from the below of grassroots, local, transnational and multi-sited coalitions, projects and activisms. Attuned to the contemporary resurgence of migrant-led and migration-related movements, and anti-racist activism, the series builds on work carried out at the critical margins of migration studies to evaluate the 'border industrial complex' and its fall-outs, build a decolonial perspective on global migration flows and critically reassess the link between (im)migration, citizenship and belonging in the cross-border future.

**Series Editors:**



**Titles in the Series:**

*Race in Post-racial Europe: An Intersectional Analysis*, Stefanie C. Boulila
*Contemporary Boat Migration: Data, Geopolitics, and Discourses*, Edited by Elaine Burroughs and Kira Williams
*Reproducing Refugees: Photographia of a Crisis*, Anna Carastathis and Myrto Tsilimpounidi
*Disavowing Asylum: Documenting Ireland's Asylum Industrial Complex*, Ronit Lentin and Vukašin Nedeljković
*Convivencia: Urban Space and Migration in a Small Catalan Town*, Martin Lundsteen
*The Migration Mobile: Border Dissidence, Sociotechnical Resistance, and the Construction of Irregularized Migrants*, Edited by Vasilis Galis, Martin Bak Jørgensen and Marie Sandberg
*The Politics of Alterity: France, and Her Others*, Sarah Mazouz and Translated by Rachel Gomme
*The Other of Climate Change: Racial Futurism, Migration, Humanism*, Andrew Baldwin

# The Other of Climate Change

## Racial Futurism, Migration, Humanism

Andrew Baldwin

ROWMAN & LITTLEFIELD
*Lanham • Boulder • New York • London*

Published by Rowman & Littlefield
An imprint of The Rowman & Littlefield Publishing Group, Inc.
4501 Forbes Boulevard, Suite 200, Lanham, Maryland 20706
www.rowman.com

86-90 Paul Street, London EC2A 4NE



**British Library Cataloguing in Publication Information Available**

**Library of Congress Cataloging-in-Publication Data**

Library of Congress Cataloging-in-Publication DataNames: Baldwin, Andrew, 1970- author.
Title: The other of climate change : racial futurism, migration, humanism / Andrew Baldwin.
Description: Lanham : Rowman & Littlefield Publishers, 2022. | Series: Challenging migration studies | Includes bibliographical references and index. | Summary: "Offers readers an alternative way of conceptualising humanism in relation to global change, one that draws in particular from black studies as opposed to one located in the ontological fold of European humanism"-- Provided by publisher.
Identifiers: LCCN 2022015270 (print) | LCCN 2022015271 (ebook) | ISBN 9781786614506 (cloth) | ISBN 9781538196489 (paper) | ISBN 9781786614513 (ebook)
Subjects: LCSH: Humanism. | Emigration and immigration--Political aspects. | Climate and civilization. | Climatic changes--Political aspects. | Forecasting.
Classification: LCC B821 .B265 2022 (print) | LCC B821 (ebook) | DDC 144--dc23/eng/20220708
LC record available at https://lccn.loc.gov/2022015270
LC ebook record available at https://lccn.loc.gov/2022015271

™ The paper used in this publication meets the minimum requirements of American National Standard for Information Sciences—Permanence of Paper for Printed Library Materials, ANSI/NISO Z39.48-1992.

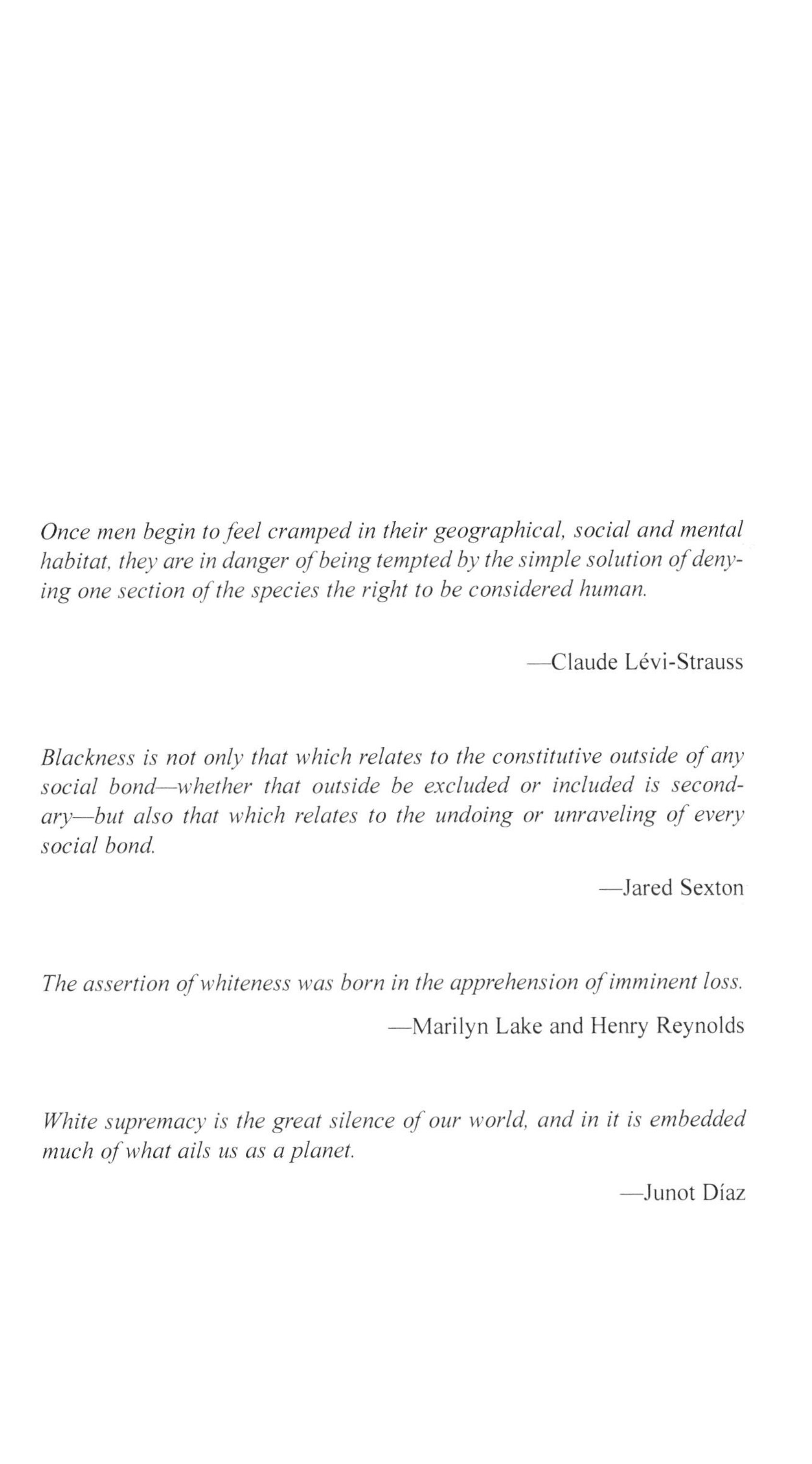

*Once men begin to feel cramped in their geographical, social and mental habitat, they are in danger of being tempted by the simple solution of denying one section of the species the right to be considered human.*

—Claude Lévi-Strauss

*Blackness is not only that which relates to the constitutive outside of any social bond—whether that outside be excluded or included is secondary—but also that which relates to the undoing or unraveling of every social bond.*

—Jared Sexton

*The assertion of whiteness was born in the apprehension of imminent loss.*

—Marilyn Lake and Henry Reynolds

*White supremacy is the great silence of our world, and in it is embedded much of what ails us as a planet.*

—Junot Díaz

# Contents

# Acknowledgements

Writing this book would not have been possible without the generosity, insight and support of countless people, not least my colleagues in the Department of Geography at Durham University. I thank each and every one of them. They include Ben Anderson, Penelope Anthias, Louise Amoore, Gavin Bridge, Harriet Bulkeley, Rachel Colls, Nick Cox, Mike Crang, Jonny Darling, Nicky Gregson, Paul Harrison, Elizabeth Johnson, Sarah Knuth, Karen Lai, Paul Langley, Jessi Lehman, Noam Lesham, Gordon MacLeod, Colin McFarlane, Siobhán McGrath, Lauren Martin, Cheryl McEwan, Aya Nassar, Leonie Newhouse, Marijn Nieuwenhuis, Joe Painter, Marcus Power, Anna Secor, Jeremy Schmidt, Phil Steinberg and Helen Wilson. I am especially grateful to Chris Orton for preparing the images that appear in chapters 4 and 5. I also wish to acknowledge former Durham Geography colleagues who also played an important part in shaping my thinking. They include Oliver Belcher, Louise Bracken, Angharad Closs-Stephens, Kate Coddington, Sarah Curtis, Stuart Elden, Chris Harker, Matthew Kearnes, Jesse Proudfoot, Ruth Raynor, Divya Tolia-Kelly and Joe Williams. The work has also benefited immeasurably from conversations with Ritwika Basu, Matilda Fitzmaurice, Jahed Mohammed Huq, Ruth Machem, Ingrid Medby, Naznin Nasir, Erwin Nugraha, Andrew Telford and Leonie Tuitjer. I also wish to thank all the undergraduate students with whom I have had the pleasure of working over the years. Finally, generous support provided by the Durham University Institute of Advanced Study and the Institute for Hazard, Risk and Resilience afforded me time and space to think and write when the ideas in this book first began to take shape.

From the period between 2010 and 2015, I had the great privilege of leading a major research network funded by the European Commission called *COST Action IS1101 Climate Change and Migration: Knowledge, Law and Policy, and Theory*. Over the course of the Action, I came into contact with the most unlikely group of thinkers, many of whom helped shape the arguments in this text. I thank them all immensely, especially Giovanni Bettini and Marco Armiero for their unwavering encouragement and friendship. I am also grateful to Jane Hogg, Catherine Alexander and Julie Dobson. Special thanks are also due to Helen Adams, Ingrid Boas, Kooj Chuhan, Dug Cubie, Carol Farbotko, François Gemenne, Saleemul Huq, Dominic Kniveton, Dimitra Manou, Robert McLeman, Anja Mihr, Paola Minoia, Sarah Nash, Calum Nicholson, Angela Oels, Lennart Olsson, Alex Randall, Julian Reid, Katherine Russo, Jeanette Schade, Teresa Thorp and Koko Warner.

From other corners of the academy, I am additionally grateful to Sonja Ayeb-Karlsson, Laura Cameron, Noel Castree, David Chandler, Nigel Clark, Mike Collyer, Simon Dalby, Jennifer Derr, Christiane Fröhlich, Ethemcan Turhan, Gaia Giuliani, Kevin Grove, Uma Kothari, Audrey Kobayashi, Jamie Linton, Uwe Lübken, Feisal Rahman, Delf Rothe, Arun Saldanha, Jürgen Scheffran, Anna Stanley, Eric Swyngedouw, and Kathryn Yusoff.

Many of the ideas that appear in this book first saw the light of day in various workshops, conferences and colloquia over the past decade. I extend my thanks to all who sat through those events while I stumbled my way through early thinking on the intersections of race, migration and climate change. Much of what appears in the book is the result of conversations had during those events. I am especially grateful to seminar audiences at University College London, University of Edinburgh, Jahangirnagar University, University of Dhaka, Kings College London, Lund University, Oxford University, Ludwig Maximilian University of Munich, Sussex University, York University (Toronto) and the University of Warwick. A version of chapter 3, "World White Order," was first presented as the keynote lecture at the "Beyond Climate Migration" workshop hosted in 2017 by the Istanbul Policy Center at Sabancı University, and a version of chapter 4, "From Determinism to Complexity," was given as the keynote lecture at the European Society for Environmental History Biennial Conference hosted by the University of Zagreb in 2017. Some elements of the writing were first published as "Climate Change, Migration, and the Crisis of Humanism," in *WIRES Climate Change* 8(3) in 2017. Parts of chapter 2 first appeared as "Racialisation and the Figure of the Climate Change Migrant," in *Environment and Planning A* 45(6) in 2013, and parts of chapter 5 were first published as "Premediation and White Affect: Climate Change and Migration in Critical Perspective," in *Transactions of the Institute of British Geographers* 41(1) in 2016. Parts of chapter 6 appeared originally in "Resilience and Race, or Climate Change and the Uninsurable Migrant: towards an Anthroporacial Reading of 'Race,'" in *Resilience*: *International Policies, Practices, Discourses* in 2017.

Adele and Quinn were both very small when *The Other of Climate Change* first began to take shape. Their part in shaping this text may be unknown to them but not to me. Their exuberance, contemplation and laughter have been among my greatest sources of inspiration. In many ways, this book is for them. I reserve my deepest heartfelt thanks to them and to Tara, without whose love, support and companionship this book would never have been written.

# Preface

On 29 November 2015, in the lead up to the Paris climate summit, an estimated 50,000 people gathered in Hyde Park to take part in the London Climate March. A clear refusal of petromodernity, the event was a gathering of the demos, the people, celebrating the transformative power of collective assembly in the face of a full-on planetary crisis. Their message was simple. Reduce greenhouse gas emissions or face dire consequences. According to the BBC, approximately 2,500 similar events were staged that day in cities around the world in an unprecedented act of global solidarity, rivalled, perhaps, only by the People's Climate March in New York the previous year. Acts of popular solidarity in the name of climate change have been a commonplace over the past decade, while in 2019, when much of this book was being written, a further wave of popular activism reinvigorated the climate change movement in the form of mass student-led climate strikes, inspired by the teenaged Swedish climate change activist Greta Thunberg and Extinction Rebellion. But on this day in 2015 an unfortunate sequence of events unfolded as the London Climate March got underway.

The event organisers had agreed that the march would be led by the Wretched of the Earth, an affiliation of Indigenous and anti-racist groups, representing the frontline communities who stand to be most adversely affected by climate change, the urban and rural poor inhabiting the most precarious margins of petromodernity. It is a truism in the moral logic of climate change to say that large areas of Earth will be rendered uninhabitable because of climate change, bringing disproportionate ruin to people of colour in the Global South. Wretched of the Earth was in London that day not only to give voice to this voiceless global majority but also to make visible what it considered to be the colonial and racist origins of climate change. But as the crowds assembled in Hyde Park that morning, Wretched of the Earth found that its symbolic position at the head of the march was soon usurped by a group of people in giraffe costume who had come to occupy the front of the procession. When Wretched of the Earth complained to the event organisers that the symbolism of displacing the voiceless Black and brown majority with people in giraffe costume repeated the colonial and racial violence that lies at the heart of the climate crisis, they were told that messages of colonialism and racism did not quite align with the celebratory and familial tone of the demonstration. Better that the victims of climate change remain victims and not politicise climate change by agitating for colonial redress. Wretched of the Earth then responded with an impromptu sit-in that impeded the procession, an act of defiance that drew scorn from many of the mostly white, middle-class demonstrators.[1]

Much can be said about this unfortunate sequence of events. It reminds us that the occupancy of public space is contingent, unequal and provisional and that the politics of climate change are irreducible to the binary logic of belief and denialism. But it also reflects a discomforting truth. As Ann Stoler once put it, "Racial thinking harnesses itself to varied progressive projects and shapes the social taxonomies defining who will be excluded from them."[2] For many who showed their solidarity that day for an oil-less future, climate change is a global phenomenon that transcends the particularities of nation, race, place and identity. It may well unleash a slow violence upon those least responsible for it; the injustices of climate change may well be *particular*. But the stranglehold that fossil fuels have over our contemporary political economy demands *universal* solidarity if it stands any chance of being loosened. For groups like the Wretched of the Earth, however, the political economy of fossil fuels is itself founded on racism and colonial violence such that to oppose oneself to climate change must entail not simply the rejection of fossil fuels but the relations of power that allow them to flow. What this brief but important episode reveals is that such relations are shaped not only by the abstract logics of capitalism but also in equal measure by various shades of racism. It reminds us that even while climate change is often imagined as a site for galvanising progressive social transformation, exemplified, for example, in Naomi Klein's popular book about climate change, *This Changes Everything*, the political discourse of climate change does not transcend all the inconvenient raced, sexed and gendered truths that mark political and economic life in our tumultuous present, a point made by Klein herself.[3] And it reminds us that even while solutions to climate change remain firmly within the spheres of technoscience and sociotechnical innovation, climate change as a site of and for politics, as a geophysical phenomenon, as a problem of capitalism and as a universalised crisis remains within the orbit of race thinking. The events that took place in London that day suggest that not everyone can seamlessly occupy the universal discourse of climate change just as it tells us that the political discourse of climate change is intimately bound up with race. The intimacies of race and climate change were once again made all too apparent in early 2020 after it came to light that Ugandan climate activist Vanessa Nakate had been cropped from an *Associated Press* photograph of herself and four white, female climate activists, including Greta Thunberg, taken at the World Economic Forum in Davos. At a moment when the Black Lives Matter social movement had brought about a new level of global race consciousness, especially within the otherwise race-blind spheres of institutionalised white humanism and liberalism, incidents like these reflect all too clearly the ease with which race is, literally, airbrushed from the political discourse on climate change.

This is a book about climate change and race. To be clear, though, it is not a book that examines how the impacts of climate change are disproportionately experienced by the Black and brown majority world. Rather, it is one about the racialisation of a specific *political* relation, that between climate change and migration. Racialisation is a complex political process. It boils down, however, to the idea that race is not an objective category but reified through forms of knowledge, such as science and law, that differentiate people by attributing meaning to signs of difference, like skin colour or language. Steve Garner describes racialisation as the process by which race "becomes meaningful" in a particular context.[4] Audrey Kobayashi and Linda Peake describe it as the "material processes and the ideological consequences of the construction of 'race' as a means of differentiating, and valuing, 'white' people above those of color."[5] Both definitions emphasise how racialisation is about the inscription of racial meaning onto people in the service of racism. But they also inform us that racialisation is geographically specific, meaning that the mechanisms by which race is reified—its scale of application, for example, or the forms of knowledge it mobilises—are different from place to place. Racialisation in the settler colonial context of Canada is vastly different, for example, than it is in France or South Africa. It can also operate through the scale of the city, or region, not just the nation. This book is focussed primarily on exploring how racialisation functions in relation to the space of the "international." Running throughout this book is the notion that the predominantly international discourse on climate change and migration is itself both a form of knowledge and an arrangement of institutions which together inscribe racial meaning onto groups of people through the category of the "climate change migrant" or what is oftentimes called the "climate refugee." The discourse on climate change and migration is, in this sense, a form of power that reproduces a racialised social hierarchy between white humanism and what we might call the racial other of climate change, the figure purportedly displaced by climate. *The Other of Climate Change* explains how I arrive at this position.

No doubt many who took part in the London Climate March in 2015 and who have since taken to the streets to bring about the end of fossil fuel use will have done so in the name of those who stand to be displaced by the impacts of climate change. This is an understandable and, in many ways, laudable position to take. After all, climate change is not an abstract phenomenon but a material condition that will affect the lives of people everywhere. Turning away from those who stand to be affected the worst by this condition would be the height of callousness. And yet I have written this book as an immanent critique of a genre of thought that takes precisely this position. Not out of callousness though but because the international discourse on climate change and migration often lacks any serious reflexivity about both its object and its own conditions of possibility. The result is a political discourse that trades in a vocabulary of difference but without any real consideration given to what this difference means. Throughout this book, I argue that the discourse on climate change and migration is bound up in the histories and politics of race and racism. My specific focus in *The Other of Climate Change* is the expression of a liberal humanist political discourse that coalesces in a desire for the climate migrant or climate refugee, a desire to intervene upon this figure for the betterment of planetary humanity.

The full nature of my concern is explained in detail throughout this book. Most straightforwardly though *The Other of Climate Change* challenges the whiteness of climate change and migration discourse. My point in writing this book is not, however, to discredit those who have taken to the streets, often at personal risk and sacrifice, to raise consciousness about the urgency of climate change or to pressure obstinate governments to divest from fossil fuel–based capitalism. Nor is it to discredit climate science or anyone who adopts even the vaguest political commitment to reversing anthropogenic climate change. Climate change is an epochal phenomenon the worst effects of which will only be avoided if varied political groups and affinities coalesce around a radically different political economic vision for the future. To call out the whiteness of climate change in a political context that demands broad political solidarity should therefore not be read as equivalent to arguing that climate change politics should be abandoned or that the climate denialists were right all along. To read *The Other of Climate Change* as in any way sympathetic to climate denialism would be to misread both my argument and my intention.

This book should instead be read as a form of immanent critique, by which I mean a form of critique that challenges many of the assumptions that organise the dominant political discourse on climate change and human migration, even while acknowledging the imperative that climate change and human migration be thought together. My aim here is to persuade readers that more serious consideration needs to be given to the myriad and surreptitious ways race infuses that relation. From this immanent perspective, making race and whiteness transparent within political and academic discussions about climate change and migration is ethically desirable if doing so means that actors become more self-conscious about how racism operates as a form of structural power that shapes not only the political terrains of climate change struggle but also their personal investments in or exclusions from that terrain. To be clear, though, *The Other of Climate Change* is not an indictment of any particular individuals or organisations. Nor is it a condemnation or an accusation of racism. Instead, this book should be read as an invitation to consider how knowledges about climate change and migration carry racial connotation and how, in doing so, they come to reinforce existing racial hierarchy.

My contention is that anthropogenic climate change should, thus, be conceived not only as a form of slow violence waged on the world's poor but also as a vast arrangement of scientific and political discourses, institutions and agents through which racialisation occurs. If anything, my purpose in writing this book is to contribute constructively to the realisation of a post-capitalist democratic future in which hydrocarbon extraction is massively diminished if not outright eliminated. But any such post-capitalist democratic political future is bound to fail unless fortified by an analysis of race and racism and empowered by meaningful cross-race political solidarities. This book has therefore been written as an effort to persuade those involved in the climate movement—student climate activists, scientists and researchers, policymakers—to confront how the political discourse on climate change and migration is enmeshed with the structural racisms against which Indigenous peoples, people of colour, the colonised, the enslaved, landless peasants, migrants and refugees all over the world have been struggling for hundreds of years. Race and racialisation are complex political phenomena that implicate everyone everywhere, myself included. I am not claiming to have it all figured out. I would, though, urge readers to follow this path. For it is only by first acknowledging how racism is present in the discourse that its grip, like that of fossil fuels themselves, might be loosened.

But if *The Other of Climate Change* is written from an anti-racist perspective, it is complicated by a fundamental contradiction, namely, what George Lipsitz calls "the impossibility of the anti-racist white subject."[6] This is the idea that even while white people, and here I include myself, may declare their anti-racism, such declarations often simply reaffirm whiteness rather than pose some fundamental challenge to it. Sara Ahmed calls this the "non-performativity of anti-racism": such declarations "do not do what they say."[7] This is not to suggest that white people should give up challenging the system of power from which they benefit. I raise this point simply to draw attention to the idea that there is no safe or neutral place to which one can retreat in the face of racism. The claim to neutrality is simply whiteness announcing itself. To argue that climate change is racist because of the violence it inflicts on Black and brown people the world over is not incorrect.

But nor can this ever be a neutral political declaration. It matters considerably who issues this declaration, why and to whom they issue it, and in what context. On this point it would be the height of naïveté to suggest some moral equivalence between Al Gore and rural peasants in coastal Bangladesh. When Gore says the West has a moral responsibility to stop climate change, whether we like it or not he is enacting a form of authority wholly unavailable to rural Bangladeshis. How we account for this power asymmetry is a question of vital political importance in a conjuncture marked by the converging crises of capitalism and Earth. I believe that whiteness, as a system of power, should lie at the heart of any attempt to explain this asymmetry. Indeed, I would even argue that the universal imperative to which Al Gore appeals in making his declaration is a direct appeal to whiteness. *The Other of Climate Change* is partly an attempt to account for this. However, and this is my point, the contradictory nature of my own anti-racism is implicated in this declaration.

I have tried to develop an analysis of the political discourse of migration and climate change which, I hope, readers will find useful for grasping how the discourse is structured by race, how the figure of the climate migrant acquires racial meaning and how the discourse reinforces a relation of power which values white people above those of colour. As a form of immanent critique, *The Other of Climate Change* is written in a way, I hope, will persuade readers to think carefully about the multiple ways in which discourse on climate change and migration is racialised. Along the way some readers will no doubt find some of my claims difficult to accept. Others might even take offense with the overall argument, especially white readers not normally used to thinking about how they themselves mobilise and benefit from racialised discourse. In fact, I suspect many readers would take the view that racialisation is something associated with the far right or ethnonationalism but not with the struggle to reverse the worst effects of climate change. After all, white nationalism is often synonymous with climate denialism in Western political culture, an association all too evident, for example, in the rhetorical brand of Trumpism and the newly emboldened white supremacies that have surfaced across the West over the past decade. But racialisation is more ubiquitous than this. It elevates whiteness in general and not simply its incarnations on the far right. It manifests, for example, in much left political thought just as it does liberalism.[8] It also functions epistemologically just as it organises everyday experience from seemingly mundane habits of speech and affect and to sexuality and social reproduction. Yet in spite of its ubiquity racialisation remains mostly obscured from the wider political discursive context of climate change. My hope in writing this text is that it will give readers a vocabulary for better understanding not only how race and whiteness shape the political discourse of climate change and migration but also, even more importantly, how this discourse serves to reinforce whiteness.

## NOTES

1. Alexandra Wanjiku Kelbert and Joshua Virasami, "Darkening the White Heart of the Climate Movement," *The New Internationalist*, 1 December 2015, https://newint.org/blog/guests/2015/12/01/darkening-the-white-heart-of-the-climate-movement; Wretched of the Earth, "Open Letter from the Wretched of the Earth to the Organisers of the People's Climate March of Justice and Jobs," *Dissidents*, 16 December 2015, https://blackdissidents.wordpress.com/2015/12/16/open-letter-from-the-wretched-of-the-earth-bloc-to-the-organisers-of-the-peoples-climate-march-of-justice-and-jobs/.

2. Ann Laura Stoler, *Race and the Education of Desire: Foucault's History of Sexuality and the Colonial Order of Things* (Durham, NC, and London: Duke University Press, 1995), 9.

3. Naomi Klein, "Let Them Drown: The Violence of Othering in a Warming World," *London Review of Books* 38, no. 11 (June 2016), https://www.lrb.co.uk/the-paper/v38/n11/naomi-klein/let-them-drown; Naomi Klein, *This Changes Everything: Capitalism vs. The Climate* (New York: Simon & Schuster, 2014).

4. Steve Garner, *Racisms: An Introduction* (Thousand Oaks, CA: Sage, 2010), 19.

5. Audrey Kobayashi and Linda Peake, "Racism Out of Place: Thoughts on Whiteness and Anti-Racist Geography in the New Millennium," *Annals of the American Association of Geographers* 90, no. 2 (June 2000): 393.

6. As quoted in Robyn Wiegman, "Whiteness Studies and the Paradox of Particularity," *Boundary 2* 26, no. 3 (Autumn 1999): 123.

7. Sara Ahmed, "Declarations of Whiteness: The Non-Performativity of Anti-Racism," *Borderlands* 3, no. 2 (2004), https://webarchive.nla.gov.au/awa/20050616083826/http://www.borderlandsejournal.adelaide.edu.au/vol3no2_2004/ahmed_declarations.htm.

8. On the racial dimensions of the liberal welfare state, see, for example, Melinda Cooper, *Family Values: Between Neoliberalism and the New Social Conservatism* (Brooklyn, NY: Zone Books, 2017); Deborah Ward, *The White Welfare State: The Racialization of US Welfare Policy* (Ann Arbor: University of Michigan Press, 2005); Domenco Losurdo, *Liberalism: A Counterhistory* (London: Verso, 2014).

# Chapter 1

# A Theory of Racial Futurism

*The Other of Climate Change* is a book about racial futurism. Racial futurism is a form of racial rule which stakes a claim on the future by constructing the racial other as that which could materialise if efforts are not taken in the present to pre-empt that future. Unlike prevailing theories of racialisation, which tend to construct the racial other in the idiom of *different-from*, the racial idiom at stake in racial futurism is that of the *yet-to-come*. Racial futurism is by necessity a reactive logic, yet one that masquerades in the language of progressive liberalism. Its terrain of operation is neither segregation nor integration but planetary circulation and adaptation. We can think of racial futurism as a form of racial rule, possibly even a form of war, that humanism turns to in order to preserve itself amidst changing global circumstances. Racial futurism is not a nostalgic desire for the past, but a logic of adaptation, the preserve of a revitalised white humanism for the future. *The Other of Climate Change* traces the logics and rationalities that liberal whiteness uses to foster the illusion of planetary control at the very moment climate change is imagined to threaten the fantasy of global racial order. It does so by examining how racial futurism structures one of the more troublesome categories in the political lexicon of climate change: the figure of the climate migrant/refugee.

Whether the rural peasant living in coastal Bangladesh exposed to cyclones, a resident of the Maldives whose livelihood is encroached upon by sea-level rise or those made homeless by evermore frequent wildfires in California or Australia, the figure of the climate migrant or climate refugee occupies an important yet enigmatic position within the discursive order of climate change. Often referred to as the "human face" of climate change, the figure of the climate migrant or climate refugee has come to signify the social dislocations resulting from the geophysical impacts of climate change.[1] In military and security discourses, the figure is often imagined as either a catalyst for future political violence or the result of climate wars.[2] For others, the figure is said to be a pending humanitarian catastrophe.[3] While for still others, it is said to play an important part in climate change adaptation.[4]

All around the world and across the spectrum of climate change political thought, there is a growing concern that climate change will inflict a profound injustice by forcing people to migrate from their homes. This injustice is encapsulated in the figure of the climate migrant or climate refugee. What makes this taken-for-granted figure so fascinating, and yet so beguiling, however, is that it occupies a controversial position at the limit of thought, on the threshold between social and environmental explanation, present and future, order and disorder, culture and nature. Over the course of this book, I endeavour to show how the discourse that brings this figure into being is constructed racially, how the figure's occupancy at the limits of space and time is racialised and how its liminality is brought into being through and, even more importantly, *for* white humanism. This book does not, however, examine directly how the impacts of climate change disproportionately affect the Black and brown majority world.[5] While full elaboration of the racialised impacts of climate change remains a vital task for critical scholarship, the focus of this book lies elsewhere. By examining how the figure of the climate migrant/refugee is racialised, my aim is to sketch the contours of what I take to be a novel and emerging form of power unfolding at the threshold of the climate change crisis. Racial futurism is the name I give to this emerging regime of power. Readers will judge whether this is a valuable interpretation. I introduce it in the pages that follow in the hope that it might open new avenues of theoretical inquiry and political engagement on questions to do with race and climate change. As critical awareness of race gains visibility within political discussions on climate change, largely but not always under the sign of "climate justice," my hope is that *The Other of Climate Change* will add meaningfully to that incipient dialogue.

From the outset, let me clarify my choice of terminology. While much effort has been made to parse the difference between a climate refugee, a climate migrant, an environmental migrant, an environmental refugee and a range of other related categories, unless otherwise specified, I use these terms more or less interchangeably. For reasons I hope will become clearer over the course of the book, in most cases I refer simply to the *figure of the climate migrant/refugee*. The question of how to label those who migrate because of climate change has been a source of perennial debate among those in the field of environmental migration.[6] As explained in more detail later, at the core of this debate, there has been prolonged disagreement about how to define those who will relocate because of climate change. Concepts like climate migrant or climate refugee are, thus, essentially contested categories, defined, paradoxically, by their lack of agreed definition.[7] Running throughout this book is the presupposition that these categories have no shared, underlying referent. There is no phenomenon called climate-induced migration or climate-related migration that is not also an immediate consequence of racial capitalism.[8]

To reflect this ambiguity, I simply conflate these categories into the singular *figure of the climate migrant/refugee*. This is for two reasons. First, to settle on a definition of the "climate migrant" or "climate refugee" would reify a relation between climate change and migration which, throughout this book, I devote considerable energy calling into question. And second, each of these categories signifies the same thing: the other of climate change.

## CLIMATE CHANGE, MIGRATION AND THE CRISIS OF HUMANISM

This book sets out to develop a political theory of the figure of the climate migrant/refugee. This is, admittedly, a tall order for such a polymorphous figure. As an object of governance, it animates domains of policymaking, from climate change adaptation policy in the United Nations Framework Convention on Climate Change to the deliberations at the United Nations Security Council on the international security dimension of climate change, from urban planning in Bangladesh to the human rights of transboundary migrants. The figure lies at the heart of climate change military planning, just as it figures centrally in climate justice discourse. It is also visible in political speech. António Guterres, Barack Obama, John Kerry, Newt Gingrich, Al Gore, Boris Johnson, Pope Francis, Prince Charles and, no doubt, many others, have all made reference to the figure of the climate migrant/refugee in their rhetoric. Images of the climate migrant/refugee circulate widely in news media. *The Guardian*, *The New York Times*, *Rolling Stone*, *Wired Magazine*, *National Geographic*, *Financial Times*, and *The Atlantic* have all run stories on either climate migration or climate refugees.[9] Both are also standard fare on university syllabi, recruitment panels and research council funding, as well as the subject of countless documentary films, books and art exhibitions. The popular cinematic and literary genre of cli-fi is replete with narratives about migration, and the climate refugee has even been the subject of a stage play.[10] How then is it possible to analyse the racial composition of a figure that appears across such varied representational terrain? How does race thinking come to define a multifarious figure like that of the climate change migrant or climate refugee? What kind of regularity organises this figure across these various genres? The answer to these questions, which I return to throughout this book, turns on the category of the human.

Dipesh Chakrabarty positions humanism at the centre of the discussion about climate change.[11] His primary concern is that climate change is putting human self-understanding to the test. Whereas the Enlightenment figure of the human was imagined as the agent of world history, capable of transforming the world through the arts, science and philosophy, Chakrabarty's concern is that climate change poses a fundamental challenge to this image of the human. "In becoming a geophysical force on the planet," he writes, "we have also developed a form of collective existence that has no ontological dimension. Our thinking about ourselves now stretches our capacity for interpretive understanding."[12] What I take this to mean is that in the historical moment of climate change, the human has become incommensurable with itself. No longer is the human a stable reference point in history, philosophy or science, if it ever was. No longer can the human be conceptualised as the agent of world history. With climate change the human must now be rethought simultaneously as the subject of modernity *and* as a geological force capable of undermining the very ecological conditions that make modern human life possible. Emerging from Chakrabarty's account, then, is the idea that as climate change takes hold of the world, the human is understood to be undergoing a profound ontological crisis.

The crisis of humanism described by Chakrabarty is of fundamental importance for understanding how racial futurism operates as a form of power on and through the figure of the climate migrant/refugee. This figure is a political category deployed by a range of different actors across a range of different political contexts for a range of different, often contradictory, purposes. What these actors all share in common, however, and here we arrive at the central thesis of this book, is that each deploys the figure of the climate migrant/refugee in a way that resolves the ontological crisis of humanism brought about by climate change. Such actors may not do so consciously or explicitly. Nor do they do so in any singular or coordinated fashion. Still, this ubiquitous other of climate change—the climate migrant or climate refugee—serves the convenient purpose of resolving the crisis of humanism that Chakrabarty observes. This, I argue, is what is at stake politically when climate change is said to be a problem of human migration. Though alongside this claim, I argue that the category of the climate migrant/refugee is not only a political category but a racial category as well. Or to be more precise, the category of the climate migrant/refugee is a signifier of Blackness, and that it acquires this connotation by virtue of its tortured relation to humanism. For at the heart of humanist thought, we also find the category of race.[13] Jean Paul Sartre was unequivocal about this in his preface to Frantz Fanon's *Wretched of the Earth*, when he states that "there is nothing more consistent than a racist humanism, since the European has only been able to become a man through creating slaves and monsters."[14] Throughout this book I document how racial futurism positions the figure of the climate migrant/refugee outside the category of the human, and how it draws the figure into being as an object of racio-political thought through the representational practices of white humanism.

By positing the figure of the climate migrant/refugee, racial futurism works to shore up "white" humanist ideology amidst the ontological crisis that climate change poses for humanism. It inscribes at the centre of the international political debate about climate change what John Hobson has called a "Eurocentric conception of world politics."[15] When the figure of the climate migrant/refugee is asserted as an object of climate change governance—the future other to be governed—racial futurism allows us to comprehend how such assertions have as much or more to do with stabilising white Euro-western humanism as they have with governing the other of climate change. Whether the migrant other of climate change is said to be the object target of humanitarianism, militarism or climate change adaptation policy, in each of these political discourses, the migrant other functions to stabilise the white Euro-western subject. Moreover, each of these areas of discourse and practice all seek to restore the human subject as the agent of planetary history at the very conjuncture in which the human and human agency are no longer the certainties they were once imagined to be. Through racial futurism, the white Euro-western subject arises, then, as a kind of universal, planetary human. A newly reinvigorated version of itself, this subject imagines itself capable of bringing order to the migratory uncertainties of climate change. If climate change poses an existential crisis for humanism, then the figure of the climate migrant/refugee does what the racial other of European thought has always done. It resolves the crisis of humanism, serving as the ground for the Euro-western subject's own self-remaking.[16]

Contrary, then, to what is often regarded as one of the most pressing geopolitical crises related to climate change, I shall therefore argue throughout *The Other of Climate Change* that the prevailing international political discourse on climate change migration is primarily one about rescuing whiteness from the climate crisis. This may strike some readers as a misguided critique. Perhaps even irresponsible. For surely the dislocations that anthropogenic climate change is bound to inflict on much of the world's population provides one of the strongest justifications for reducing greenhouse gas emissions. Is it not the case that those in the industrialised North have a moral responsibility towards those in the Global South who will be made worse off by climate change while having contributed to it the very least? Would it not undermine the credibility of climate justice activists were they to argue that the climate change migrant is a racial category? Very possibly. But nor should we assume that the signifier "climate justice" or the questions of moral responsibility are not themselves already bound up with questions of race and racism. After all, it is precisely the uneven racialised geography of North and South, a geography central to the formation of climate justice, that W. E. B. Du Bois captured in his prescient claim that "the problem of the twentieth century is the problem of the color-line,—the relation of the darker to the lighter races of men in Asia and Africa, in America and the islands of the sea."[17]

A few simple examples clarify why I believe an analytics of racial futurism for the figure of the climate migrant/refugee is warranted. In their book, *Rising Tides: Climate Refugees in the Twenty-First Century*, published by Indiana University Press, John Wennersten and Denise Robbins are unequivocal that the category climate refugees refers to Black and brown people from the Global South.[18] They write, for example, that "it is primarily people in poor developing countries that have high population density, marginal food stores, health problems, and political instability who will become environmental refugees."[19] "It is a safe bet," they say, "that they [climate refugees] will lap up on the shores of prosperous developed Western nations,"[20] and, later, they implore us to "examine new methods of immigration and refugee absorption that spread new aliens throughout new environments to prevent the development of refugee camps, slums, and ethnic ghettos of people of colour in primarily Caucasian landscapes."[21] Wennersten and Robbins clearly mobilise a racist geopolitical imaginary that distinguishes between (White) Europe and the West from the Black and brown majority world. By their reasoning, climate change is not simply a pending migration crisis, but a racial crisis. Or to take a different example, consider the photographic collection *Climate Refugees*, produced by Collectif Argos. Of the 175 photographs featured in the collection, 128 of these depict Black and brown people, 7 depict white people, and 1 image is of person whose "race" is undiscernible.[22] (The latter should alert us to the fact that reading race off the body is always fraught.) Or do a simple online browser search for "climate refugee." What turns up are dozens of images, all of which reveal a similarly racist geopolitical imaginary; large groups of Black and brown human beings fording bodies of water or trekking across desiccated landscapes. Far from misguided, then, attending to the work race does in relation to the figure of the climate migrant/refugee is simply to respond to what is already evident in the discourse.

But what exactly does "race" do? What kind of work does it perform in relation to the genre of humanism at stake in the global climate crisis? Throughout this book, I take race to be an effect of white humanism. Race is a political technology white humanism must invent in order to naturalise Blackness outside the category of the human. "Racisms," as Stuart Hall had once famously put it, "also dehistoricize—translating historically specific structures into the timeless language of nature."[23] As one of nature's collateral concepts, race is the category racists use to naturalise social relations under capitalism.[24] It legitimates uneven, violent constructions of difference by explaining these as a function of nature as opposed to social forces, such as history and capital, that structure the life chances and everyday decisions of those subject to racist violence. This is why we should find the figure of the climate migrant/refugee so troubling; it translates all manner of peoples' everyday vulnerabilities, emergencies and disasters historically produced through the violence of colonialism, enslavement, imperialism, patriarchy and racial capitalism into the language of nature. This occurs because, within the prevailing discourse on climate change and migration, the figure of the climate migrant/refugee is represented primarily as a function of climate change—human mobilities precipitated by, for example, sea-level rise, extreme weather, climatic variability—yet without properly taking into account the way climate change itself is the culmination of these relations of violence. In this way, climate change functions as a proxy for nature, a human-made geophysical phenomenon shorn of its historical substance.[25] Part of my argument in *The Other of Climate Change*, then, is that by attending to the way in which the white Euro-western subject mobilises climate change to naturalise the figure of the climate migrant/refugee through race, by attending to the figure's racialisation, we might come to appreciate how whiteness itself is adapting to the existential threat that climate change poses to humanism.

My argument is indebted to Geoff Mann and Joel Wainwright's compelling thesis *Climate Leviathan*.[26] These authors make the case that comprehending capitalism in relation to climate change means attending to the "adaptation of the political."[27] This means rigorously analysing how capitalist political economy is itself adapting to the instabilities of climate change. They define the political very precisely as that which "defines a relation tout court: the relationship between the dominant and the dominated," the very "ground upon which the relation between the dominant and dominated is worked out."[28] I argue throughout this book that the assemblage of discourses, institutions, practices, affects and social relations that comprise the encompassing relation between climate change and migration is precisely this ground. If we want to understand something about how the political adapts to climate change, if we want to understand something about the way in which "the ground upon which the relation between dominant and dominated is worked out" amidst the instabilities of climate change, then we should turn our attention to the climate-migration relation, for it is here, buried deep within that relation, that the terms of domination and subordination specific to climate change are being crafted. The figure of the climate migrant/refugee, I suggest, is precisely the ground on which the political is adapting to climate change. This is not to challenge Mann and Wainwright's important analysis. As they rightly argue, "There is no moral alternative to giving much greater attention to climate refugees."[29] It is simply to position the question of migration and mobility squarely at the centre of their discussion about the adaptation of the political. If human survival on a warming planet involves large numbers of people having to relocate as a matter of survival, then the very terms by which such migration is understood and, thus, governed in relation to planetary habitability should sit at the heart of any analysis that seeks to diagnose how the political adapts to climate change. If anything, *The Other of Climate Change* is simply one answer to Mann and Wainwright's call for "a critical elaboration of these terms."[30]

Over the course of this book, I critically elaborate these terms by attending to racial futurism and the logics of anti-Blackness that animate it. The risk of not doing so is that political analyses of climate change may neglect to consider how race thinking becomes harnessed to the myriad projects of mitigating and adapting to climate change. The result of this neglect, moreover, is that the political discourse of climate change risks becoming a site for the unwitting rearticulation of racial hierarchy, as opposed to a site over which race or, more precisely, whiteness as a specific structure of power, is contested and resisted. I am not suggesting that I have this all completely worked out, and there are numerous aspects of the discourse on climate change and migration not covered in this book, most notably law and human rights, which are dealt with only minimally in chapter 2. Nor does the book undertake any real sustained engagement with specific *in situ* political struggles which converge on the issues of climate change and migration and/or displacement. My neglect of these considerations may come back to haunt me. Nor is migration the only domain of politics through which to interpret race and climate change.[31] Still, my argument makes the case that contained within the political discourse on climate change and migration is a politics of race. Attending to that politics is, in my view, a matter of pressing urgency if we are to avoid repeating the violence and bloodshed that mark the history of humanism.

My thinking about racial futurism is also deeply indebted to the work of Giovanni Bettini. Especially noteworthy is his incisive analysis of the discourse's numerous apocalyptic narratives about which he makes two pertinent arguments.[32] The first is that apocalyptic narratives of climate change and migration call attention to the way in which the relation's "political significance" is intimately bound up with its "ontological status." Although I use the vocabulary of apocalypse only sparingly, my conception of racial futurism owes much to Bettini's claim that the political dimensions of the discourse are primarily ontological. Second, Bettini argues that climate change and migration discourse is an effect of signification which produces a surplus of meaning. By this he means that the discourse has no stable meaning but is instead concocted from several vocabularies, all of which problematise climate change as a migration crisis but for different reasons. Militaries and security agencies construct it as a catalyst for war or political violence; humanitarians frame it as a pending humanitarian crisis, while development agencies may construct it as a problem of poverty. In this sense, there is widespread agreement that climate change is a pending migration crisis, even if actors disagree on the terms used to describe it and what it means. The fact that the relation has no inherent meaning, only a surplus of meanings, is precisely the reason why migration has become normalised as a condition of climate change. On this crucial point I concur wholeheartedly with Bettini.

But even as *The Other of Climate Change* shares much in common with Bettini's important work, my argument diverges from his in two important respects. First, it works across the gap between race and the political. So, while Bettini is entirely right to claim that the discourse produces a surplus of meaning, I argue that this surplus coalesces around white humanism. That is to say, all of these "apocalyptic narratives" share in common an unstated affinity for whiteness. Drawing from Martin Hajer's concept of "discourse coalitions," Bettini argues that these discourses (e.g., security, humanitarian and development) "converge . . . not so much because of some trait inherent to the narrative" but because these discourses have some shared affinity. As he puts it, "A certain narrative resonates with those discourses and 'sounds right.'"[33] But what kind of aurality is this? How can a discourse coalition hold together on the basis of a narrative resonance that sounds right? I argue that this feeling, or aurality, of "sounding right" *is* whiteness. Whiteness can, of course, be conceived as an identity and political subjectivity, but it is foremost a form of power that functions through a shared sensibility that precedes social differentiation. Second, my argument extends Bettini's claim that the discourse "eras[es] the room for an emancipatory, democratic politics of C–M (climate-migration)." Bettini is right to claim that the discourse on climate change and migration dulls the political imaginary, reducing the future to a migration crisis that must be confronted and managed. I argue in chapter 5, however, that this diminished political imaginary should be construed racially as white affect. This is a white political imaginary which functions to ensure the international political discourse on climate change remains in thrall to a universal white humanism. It diminishes the creative affordances of the future, foreclosing the future as a political resource for remaking the world in ways that not only challenge but also undo white humanism. If Earth is desperately in need of a reboot, racial futurism is the form of rule white humanism turns to in order to ensure that it remains the central "operating system" of whatever comes next.[34]

## A GENEALOGY OF THE FIGURE OF THE CLIMATE MIGRANT/REFUGEE

How can we account for the emergence of the figure of the climate migrant/refugee as an object of international governmental concern? What are the figure's historical conditions of possibility? Early evidence for the figure of the climate migrant/refugee within climate science can be found in the First Assessment Report of the Intergovernmental Panel on Climate Change (IPCC) published in 1990.[35] Prior to this, the figure of the environmental refugee had been popularised in a famous report published by the United Nations Environment Programme in 1985 titled simply *Environmental Refugees*.[36] Richard Black attributes the first usage to Lester Brown at the Worldwatch Institute in the 1970s.[37] However, the figure of environmental migrant has an extensive genealogy, traceable at least to nineteenth- and early twentieth-century environmental determinism where climate and migration occupy the ideological heartland of white mythology's fascination with race, evolution and civilisation.[38] I examine some of this genealogy in chapter 4. No doubt the figure can also be found in earlier attempts to naturalise the long arc of Western history: the fall of Rome, for example, or the tumultuous period of the French Revolution.

It was not, however, until 2003, with the publication of a report commissioned by the Pentagon titled *An Abrupt Climate Change Scenario and Its Implications for United States National Security*, that the figure of climate migrant/refugee would become the object of political and economic calculation it is today.[39] Up until that point, the term of choice was "environmental refugee." The Pentagon report merely reframed it within the vocabulary of climate change, thereby newly associating climate change and its migrant other as novel catalysts of political violence and humanitarian emergency. A similar framing is also evident in the United Nations Security Council debate on climate change in 2007.[40] It would not be until 2010, however, that migration would make its debut as a formal object of international climate change governance in the Cancún Adaptation Framework (CAF). This is an important milestone in the invention of the concept. Although heralded at the time as a major advance in climate adaptation policy, paragraph 14(f), which called upon signatories to "enhance understanding, coordination and cooperation with regard to climate change induced displacement, migration and planned relocation," served to anchor the relation between climate change and migration in the United Nations Framework Convention on Climate Change (UNFCCC) more so than it did to inaugurate a formal policy.[41] Or more accurately, the policy is one that would henceforth institutionalise discursive ambiguity about the object into the legal framework of international climate change governance. This reference to migration in the CAF continues to have lasting consequences. For what paragraph 14(f) enshrined was a search for knowledge about a relation that can only properly be described as evasive. What makes paragraph 14(f) so significant, moreover, is that it embeds within international climate law a long-standing tension in modern epistemology between knowledge of "nature" and that of "culture," albeit a tension that is, like the category of race, violently enacted on the bodies of actual people. While ostensibly a request for science about the social impacts of climate change, the CAF is more pernicious because it opens up the possibility for lawmakers to differentiate people on the basis of the geophysical impacts of climate change. This same epistemological tension resurfaces again and again in the discourse on climate change and migration and, as will be made evident momentarily, supplies the basis for my argument that the discourse is a project of racial futurism.

Subsequent to its inclusion in the CAF, the climate-migration relation would go on to become further institutionalised in various international legal settings and fora. The more prominent of these came to be known as the Nansen Initiative, a global consultation sponsored by the governments of Norway and Switzerland on legal protections for transboundary migration resulting from disasters, including climate change. In operation from 2015 to 2017, the Nansen Initiative was among the outcomes of the Nansen Conference on Climate Change and Displacement held in Oslo in 2011, drawing its name from Fridtjof Nansen, the Norwegian diplomat and scientist who became the League of Nations' first High Commissioner for Refugees in 1920.[42] The climate-migration relation is also embedded in the Warsaw International Mechanism on Loss and Damage (WIM), which now represents the third pillar of international climate change governance, alongside mitigation and adaptation. In 2015, the Paris Agreement mandated a Task Force on Displacement (TFD) under the auspices of the WIM, and in 2018 the TFD made its first recommendations to the WIM Executive Committee, including the recommendation to "enhance research, data collection, risk analysis, and sharing of information, to better map, understand and manage human mobility related to the adverse impacts of climate change." More recently the Global Compact for Migration, endorsed by the United Nations General Assembly in 2018, also included a set of non-binding commitments in reference to climate change. Among others, it calls on signatories to "strengthen joint analysis and sharing of information to better map, understand, predict and address migration movements." The International Organization for Migration (IOM) has also been a driving force behind the climate-migration relation, as has the Office of the United Nations High Commissioner for Refugees.[43]

References to the relation are evident in numerous other contexts, most notably in popular media, in military planning and in national adaptation planning. But it is the matrix of international institutions, backed by the full weight of the international community, and the capitalist political economy they uphold, that has been undeniably significant in securing the climate-migration relation as an object of thought, an object of climate change governance, and, most significantly, as a racial project. This significance is partly due to the way these institutions seem collectively preoccupied with the uncertainty that surrounds the figure but also with the way they seek to go about constructing it through the production of knowledge. Thus, at stake in the institutionalisation of the figure of climate migrant/refugee, at this particular conjuncture, is the power to authorise the invention of a new form of knowledge, one that can be used to differentiate people on the basis of a specific, historically produced form of nature: climate change. This is a procedure which first assumes the objectivity of the climate migrant/refugee and then attempts to explain its existence using the language of nature. The parallel here with scientific racism is not incidental. Scientific racism first assumed the existence of race which it then sought to define in biological or genetic terms, themselves both forms of nature. I would argue that something similar is at stake in the knowledge about the figure of the climate migrant/refugee. These are forms of knowledge which seek to locate the figure's difference in climate or, more accurately, in the pending complex emergency of climate change.

Another potent source of invention, and also one examined more thoroughly later, concerns the use of quantitative estimation. The World Bank, yet another powerful international institution, estimates that "by 2050—in just three regions—climate change could force more than 143 million people to move within their countries."[44] This estimate is reminiscent of one made famous in the mid-1990s by Norman Myers who claimed that 150 million people will be forced to migrate due to the impacts of climate change by 2050.[45] The IPCC cited Myers's estimate as the authoritative number in one of its early scientific assessments in 1995. Myers's updated estimate—200 million by 2050—would go on to become the authoritative estimate in numerous influential publications and reports, including the popular U.K. *Stern Review on the Economics of Climate Change* in 2006.[46] It now circulates as an established truth, even as Myers himself has questioned its veracity,[47] and in 2007, the British charity Christian Aid had put this figure at 1 billion, more than one-tenth of the world population by current projections.[48] These are staggering numbers. The trouble with such quantitative estimates, however, is that even as they dramatise the urgency of climate change, they nevertheless contribute to the reification of a contested phenomenon, and in so doing, they end up masking how they themselves are complicit in the project of racial futurism.

## EPISTEMIC DISJUNCTURE OF CLIMATE CHANGE AND MIGRATION

At the heart of the epistemic disjuncture of climate change and migration is the assumption that climate change causes migration. This assumption has been so thoroughly debunked that expert opinion is now virtually unanimous that causation is impossible to ascertain.[49] It is hardly radical to insist on this point. Nor is it the basis on which to construct a radical politics of climate change. The IPCC, the highest international authority on climate change science, argues, for example, that "the dynamics of the interaction of mobility with climate change are multifaceted and direct causation is difficult to establish."[50] Simple historical analysis would quickly confirm this authoritative view inasmuch as the historical contingencies of place are as significant for explaining migration as environmental change. Thus, it is widely acknowledged in migration studies, and paradoxically even in the field of environmental migration itself, that migration is complex, the result of a range of political and economic circumstances, never just the "environment," "climate" or "nature." To suggest, then, that climate change supersedes the myriad other reasons for which people migrate—insecure land tenure, exploitative or non-existent labour markets, genealogies of violence, for instance—is therefore misleading, tantamount to what is now sometimes referred to as neo-environmental determinism or climate reductionism.[51] Yet, perhaps not surprisingly, this is precisely what is implied in much of the knowledge produced about the relation between *climate change* and migration: climate change is *always* elevated as the primary driver, or trigger, of migration, even while it is said to be just one of the many migration drivers. In chapter 4, I argue that the discourse on climate change and migration, while characterised by the explicit rejection of environmental determinism, is actually best understood as a revival of a weak environmental determinism.

The absence of causal certainty between climate change and migration has led Calum Nicholson to call this as an "equivocal' relation," one characterised by an irresolvable contradiction between its epistemological and ethical content.[52] What Nicholson means is that knowledge of the relationship between migration and climate change remains insufficient for intervening ethically in the relation itself. Simply put, it means that one cannot intervene ethically upon a crisis one is unable to grasp. For having only a *sense* that climate change is a mounting migration crisis raises troubling ethical questions, because sensing a crisis is not always sufficient ground for ethical action. In some cases a sensed crisis may well be sufficient grounds for intervention. When we sense a child is in danger, for example, we take action to prevent the danger from materialising. Or in the case of climate change, we have a high degree of certainty that climate change is a surety, so we seek to mitigate it, even if how that is to be done remains contested. But resettling thousands, possibly millions, of people, or deciding which groups of people are best suited to adaptive migration, because we *sense* those groups are at risk of future harm, is an ethical dilemma on a different order of magnitude. It is precisely because these ethical considerations arise that we need to scrutinise how and for whom knowledge about migration and climate change is put to use in the world.

The academic field of environmental migration approaches this equivocal problem with regular calls for more and better research, insisting that only an objective understanding of the phenomenon can provide sufficient justification for ethical intervention.[53] The same holds for the UNFCCC and the Global Compact for Migration discussed earlier. Both authorise their signatories to gather more knowledge about the interactions of migration and climate change. My contention, however, is that no amount of knowledge will ever suffice to resolve this ethical quandary. This is because any attempt to explain migration as a function of climate change will always end up subordinating historical explanation. Thus, the inherent danger of such equivocal phenomena is that, unless subject to rigorous scrutiny, they risk decoupling power and accountability.[54] A vivid example of this occurs in relation to the claim that the Syrian civil war is the result of climate change. In one notable study, it is argued that the war was the indirect result of a climate change–induced drought which drove large-scale rural to urban internal migration and which in turn led to the civil uprisings and eventually war.[55] The study is now immensely popular because it substantiates the apparent "missing link" between climate change, migration and violent conflict. But after carefully reviewing this argument, Jan Selby, Omar Dahi, Christiane Fröhlich and Mike Hulme found "no good evidence to conclude that global climate change–related drought in Syria was a contributory causal factor in the country's civil war." What they found instead was that the link between the drought and migration "was, in all probability, more caused by economic liberalisation."[56] What this example illustrates is that, unless subject to critical scrutiny, the notion that climate change is a primary contributing factor to migration can be used to misrepresent the underlying political and economic causes of migration or, indeed, war. It is not difficult to see why powerful institutions, such as Western militaries, international organisations or national governments, with an interest in upholding capitalist political economy will mobilise climate change to explain what are inherently contestable and historical phenomena. Doing so allows these institutions to draw critical attention away from the very mechanisms that guarantee their power, including, in this example, capital and the construction of scientific knowledge.

*The Other of Climate Change*, thus, challenges one of the foundational tenets of the academic field of environmental migration research, that the environment, including climate change, can explain migration. On the contrary, it rests on the idea that no amount of knowledge that explains migration as a function of climate change will ever be sufficient to supersede the underlying historical geographical dimensions of migration. The destructive legacies of plantation slavery and the ongoing devastations of racial capitalism saturate our present in ways that make them all but impossible to be explained away by climate.[57] Climate change is, of course, a real material phenomenon. But it is not a crisis of greenhouse gas emissions, deforestation, automobility or consumptive desire. These are merely manifestations of racial capitalism's perpetual violence. Throughout *The Other of Climate Change* the relation between climate change and migration is not, therefore, taken to be an objective condition. Nor does it assume that empirical knowledge about this relation is somehow politically inert. Instead, the argument rests on the claim that it is precisely this epistemic disjuncture—*the very impossibility of the figure of the climate migrant/refugee*—which, paradoxically, defines the climate-migration relation.

In emphasising this disjuncture, however, my point is not to dismiss the need to theorise migration in the context of climate change. Nor is my claim that climate change does not matter. Like Mann and Wainwright, I endorse the need for "a robust political language defending the right of people to migrate in anticipation of climate change."[58] Instead, I wish to make a different point: this epistemic disjuncture has a very powerful structuring effect on the overall temporality of "climate change and migration" discourse. For it is precisely the fact climate change can never be said to cause migration that structures *how* the relation can be imagined. That is, in the absence of causal certainty, those who construct climate change as a problem of migration are only able to do so in the grammar of the future perfect—that which could happen in the future. Thus, the discourse on climate change and migration is written overwhelmingly in the future perfect tense.[59] Migration, we are regularly told, is a phenomenon that either *will* or *may be* amplified under conditions of climate change. Moreover, any attempt to identify climate-induced migration *in the present* is always met with the inconvenient truth of absent causality.

Examples of the future perfect in the discourse are plentiful. The most common is the use of quantitative prediction, such as Norman Myers's mentioned above. Others include forms of modelling, such as statistical regression and correlation analysis.[60] The use of scenario planning is also widespread, for example, in military and policy texts, while the World Bank has developed future migration scenarios in combination with gravity modelling, development pathways and climate impact modelling.[61] Arguably the most succinct expression of this future-perfect grammar I have come across, however, comes from the international relations (IR) scholar Gregory White, who writes that "*the evidence is abundant*: a combination of rising sea levels, increasing temperatures, and changing precipitation patterns *will likely* affect migration patterns in the decades to come."[62] White is, of course, not incorrect in making this claim. It is entirely reasonable to assume that climate change will likely generate a range of effects, including migration. But there is also abundant evidence that numerous other phenomena will also likely affect migration in the decades to come. Shifting patterns of economic development will place new demands on labour migration. Shifting distributions of geopolitical power, new wars, migration bans, camps, pandemics, and large-scale migrant deportations from Europe and North America will also alter migration patterns. Climate change is hardly unique in this respect.

## RACIAL FUTURISM

Over the course of this book, I say more about why and how racial futurism matters. At this point, though, we can draw several general observations. The first is that anti-Blackness is central to racial futurism. Borrowing from Frank Wilderson, I take anti-Blackness as *the* structural antagonism that founds humanism. That is to say, anti-Blackness is precisely that which produces 'Black' as an index of existence, or more accurately an index of a life synonymous with non-existence, death and violence. Wilderson describes Blackness as social death.[63] In a material sense, anti-Blackness is what produces the global phenomenon which Ruthie Gilmore would call the "premature death" of those racialised as Black, whether in the form of police brutality, deaths in custody, police shootings/killings, slow violence, exploitation, dispossession, genocide or enslavement.[64] We could also safely add climate change to this list, inasmuch as it is said to disproportionately affect the Black and brown majority world—a common framing in genres of climate justice. From this perspective, tracing the continuities between plantation slavery, colonial violence and climate change remains an urgent task.[65] However, it is the philosophical meaning of anti-Blackness that is most pertinent for understanding racial futurism. This anti-Blackness is the *ontological* violence directed at the figure of nothingness, the figure of non-being, which, in the ideology of humanism, is the figure of Blackness.[66] It is in this latter sense of anti-Blackness that *The Other of Climate Change* examines how racial futurism structures the figure of the climate migrant/refugee.[67] For it is precisely through its displacement onto nothingness, its manifestation *as* nothingness, that, I argue, the figure of the climate migrant/refugee functions as the racial other upon which the white Euro-western subject attempts to ground its political adaptation to climate change.[68]

To clarify, consider the epistemic disjuncture noted earlier. Scratch the surface of the discourse and it quickly becomes apparent that there is no such thing as a "climate migrant" or a "climate refugee" nor any specific instance of migration that is directly attributable to climate change. This may seem counterintuitive, but the simple reason for this is that migration can never be explained solely as the result of climate change. As with all social phenomena, migration is irreducible to any single explanatory variable and is, therefore, widely considered within the field of migration studies to be contingent upon a range of factors, never just one. It is precisely for this reason that the IPCC has argued that "the use of the term climate refugee is scientifically and legally problematic," citing in particular the legal scholar Jane McAdam who argues that the concept is "erroneous as a matter of law and conceptually inaccurate."[69] We should, therefore, take from this basic analysis the idea that the related categories of 'climate migrant' and 'climate refugee' have no objective basis in either science or law. They are in this sense categories that correspond to nothing. Chapter 2 considers how the figure of the climate migrant/refugee is coterminous with nothingness, expelled from the category of the human through its positioning outside the limits of representation. I need to be absolutely clear, however, that in taking this position I am no way attempting to diminish the material anti-Blackness of climate change. If anything, my argument is that through the habitual erasure of its historical continuity with anti-Blackness, the discourse on climate change and migration does much to obscure the material anti-Blackness of climate change. For example, by severing any meaningful connections between climate change and racial capitalism, the discourse reduces climate change to a set of geophysical impacts, thus, foreclosing Leon Sealey-Huggins's notion that climate change is really just the "apolcalyptic trajectory of racial capitalism."[70] In this sense, the discourse on climate change and migration not only reinforces whiteness, it does much to ensure that the political discourse on climate change *remains* white.

Another important consideration shaping my argument about racial futurism concerns the specific scale of my analysis. Although the figure of the climate change migration/refugee is sometimes articulated through geographies of the nation and, increasingly, the urban, I approach the figure primarily through the scale of the international. This is because the international is one of the primary geographical spaces through which the figure of the climate migrant is articulated as an object of governance. This is not surprising, given that the climate is often said to be a global public good and that climate change is a collective action problem, the solution to which many imagine will come from internationally negotiated climate change policy. In this way, it makes perfect sense to examine how the figure of the climate migrant/refugee surfaces in the space of the international. But the international itself is not an innocent category. Rather, it is increasingly recognised as a political concept bound up with the histories of European racial and colonial oppression, an insight inspired in part by Du Bois's notion of the colour line.[71] And this is important because the figure of the climate migrant/refugee is often used to signal that the climate change crisis is not simply an environmental crisis but a *global* crisis of governability, or more accurately, a crisis which, if not properly pre-empted, will materialise as an ungovernable event on a global scale. This is why the climate migrant/refugee is often figured as the "human face" of climate change.[72] Millions of people moving inconveniently around the world, whether within or across territorial boundaries, is certainly a worrying prospect for those international institutions tasked with maintaining the existing order of the world. My contention is that the climate migrant/refugee concept is invoked to bolster the performative nature of that authority as it confronts the ungovernability of climatic upheaval on a global scale.

This brings us to a third set of considerations. How can we account for this form of authority? How might we comprehend it? The racial futurism that I chart in this book shares much in common with Michel Foucault's concept of biopolitics. In many ways, racial futurism is nothing more than a means of administering life on a planetary scale. Yet racial futurism is not reducible to biopower and differs from it in several key respects. Most fundamentally, racial futurism does not conform to the biopolitical formula "making live and letting die." While it might be tempting to argue that within the framework of racial futurism the figure of the climate migrant/refugee is the figure liberalism must kill in order that it may live, racial futurism does not seem to approach the figure in quite those terms. The figure of the climate migrant/refugee is most certainly the figuration of social death, but its relation to racial futurism is neither killing nor abandonment but absorption. Much like a garden, the figure must be cultivated and sustained as a source of white renewal not an object targeted for elimination. Nor is racial futurism concerned with making live; it does not target life processes at the level of either the self or population. It is not a productive form of power but is, perhaps, best understood as adaptive. In this way, racial futurism operates more as a topology of normation. Not a normalising power that disciplines in accordance with a norm, but one that seeks to grasp the event prior to it unfolding such that it may shape the norm itself.[73] Another important difference is racial futurism's terrain of application. Whereas biopower is exercised in relation to a defined territory and population, racial futurism is exercised over what Brian Massumi would call a "proto-territory," or the pre-territorial, a kind of indeterminacy of forces without precise spatial form.[74] This is the problem space of emergence, not the bounded territoriality of the modern political imagination. Racial futurism is, thus, more akin to Massumi's ontopower than biopower. If ontopower operates on ontogenetic processes, racial futurism can be understood to operate in an ontogenetic field prior to race itself.

Racial futurism bears some resemblance to what Stephen Collier calls topologies of power[75] or the reconfiguration of government. For Collier, a topology of power is recombinatorial. It responds to moments of upheaval by recombining existing governmental techniques in novel ways when old arrangements are deemed to be outdated and, therefore, insufficient. Racial futurism is one such topology of power. It is not an old form of racism now being resurrected in the context of climate change. It is, I would argue, a genuinely new topology of racism, developing in relation to a genuinely novel set of geohistorical circumstances that did not exist empirically in the nineteenth century as they do now: climatic upheaval, new forms of computation, financialisation, new patterns of migration, shifting geopolitical fault lines, environmental degradation, population growth and sustained sociopolitical challenges to humanism. All of these present themselves in ways no prior forms of racism have ever had to confront. Racial futurism can therefore be understood as the strategic reformulation of racism amidst these upheavals. We might say that racial futurism is a political assemblage that combines racist geographical imaginations, the techniques of climate science, socioecological systems, notably resilience and adaptive capacity, the global colour line scenarios, racialisation, risk management, the new economics of labour migration, white affect, IR and more in a way that brings the figure of the climate migrant/refugee into being as an object of thought the primary purpose of which is to shore up white humanism.

But if racial futurism is recombinatory, it is not exercised by any sort of sovereign authority. There is no master subject that lies behind white humanism plotting its adaptive evolution. If whiteness is an endlessly reproducing psychodrama, racial futurism is its corresponding topology of power, reconsolidating racism at the leading edge of white reorientation to climate change. It is held together by a shared sensibility rather than by a figure of authority or a law. And it is for this reason that I am reluctant to liken racial futurism to the form of authority Mann and Wainwright call *planetary* sovereignty, even though the racial futurism I examine in *The Other of Climate Change* is planetary in reach. It is quite literally a planetary arrangement of power in the making. I can understand why Mann and Wainwright would turn to Hobbes's *Leviathan* to characterise the logic of political sovereignty they see adapting to the calamitous world of climate change. Hobbes famously conceptualised sovereignty as that which would domesticate the state of nature, a condition of war of "every man, against every man."[76] Like Hobbes's *Leviathan*, planetary sovereignty undertakes to secure itself in relation to that unfolding calamity. But as a topology of racism developing in relation to the problem space of climate change and migration, racial futurism is irreducible to sovereign authority even while it draws elements of sovereignty into its orbit.

Throughout the argument, I make periodic reference to the notion of racial nomos, a concept I borrow from Paul Gilroy who uses it to designate the spatialisation of power that upholds racial order, a kind of law or rule of race sustained by racism.[77] I make use of the racial nomos concept to help characterise precisely the way this image of ungovernability—the undoing of racial nomos in the context of climate change—provokes a similar desire for order and control akin to the political sovereign imagined by Hobbes, albeit a desire that is above all structured by anti-Blackness. My use of the nomos concept is, however, also in direct reference to the concept made famous by the Nazi jurist Carl Schmitt. In his book *The Nomos of the Earth*, Schmitt used the concept of "nomos" to designate what he saw as the planetary spatial order that European modernity imposed on Earth from the end of the fifteenth century up until the nineteenth century.[78] Nomos captures the full geographical weight of European imperialism ordered through a world geography of sovereignty. But it also captures what Schmitt saw as Europe's waning geopolitical significance by the end of the nineteenth century and its imminent replacement by a new world order in the form of American imperialism.[79] Gilroy's reformulation of nomos as *racial* nomos names what Schmitt implied but never said.

Racial nomos captures this double sense of nomos. It indexes both racial spatial order and the always-present potential that this racial order could start to unravel at any moment, that climate change may render the world ungovernable racially. For Schmitt, spatial nomos required continuous maintenance as a prerogative of that order. Political sovereignty in this sense required the continuous designation and policing of that which was said to be exceptional to itself. The inverse of Schmitt's reasoning was that without firm command over the exception, political sovereignty would come undone, a lingering fear always on the mind of the sovereign. As I have already intimated, a similar "pre-sentiment of disorder" animates the discourse on climate change migration.[80] Catastrophe and apocalypse imbue the climate change imaginary with a worrying sense of planetary destabilisation.[81] The implication is that by proliferating the movement of people around the world, climate change will deterritorialise the existing global spatial order, which will be replaced eventually by some new world order, some new racial nomos. Racial futurism should be understood as a form of power attempting to reformulate racial nomos in the face of its undoing. But racial futurism is not a power reducible to the reconsolidation of sovereignty.[82] It is, rather, a form of power struggling to reassert an order of governance in the face of its undoing. In this way, the figure of the climate migrant/refugee is important not only because it signifies the boundary between order and disorder, governability and ungovernability, between racial nomos and its antithesis, but also because it signifies the breakdown of this order of governance.

What order of governance is this? It is one, I would argue, that can be traced back to the original scene of imperial discovery when Europe first encountered its non-European other.[83] Anne McClintock describes this moment as one of deep ambivalence between order and disorder, a moment marked by the simultaneous fear of emasculation and the desire for power and control. For McClintock, what arises in this ambivalent moment is an order of governance that attempts to resolve "the dread of catastrophic boundary loss (implosion), associated with fears of impotence and infantalization" with "an excess of boundary order and fantasies of unlimited power."[84] Katherine McKittrick argues something similar. For her, it is precisely this moment of encounter with the "uninhabitable—in particular, the landmasses occupied by those who, in the fifteenth and sixteenth centuries, were unimaginable, both spatially and corporeally" that gave rise to the slave plantation.[85] Both authors describe how Europe sought to resolve this deeply unsettling encounter with non-European difference with a distinct order of racial governance. In chapters 2 and 3, I develop the idea that the figure of the climate migrant/refugee carries traces of this imperial ambivalence and the "geographic (non)location" associated with the plantation. There I argue that the figure is a liminal *racial* figure, invented to stabilise a contemporary iteration of this same order of governance—the postcolonial white international—that climate change is now threatening to unravel. As Gilroy argues, "Racial difference and racial hierarchy can be made to appear with seeming spontaneity as a stabilising force . . . to secure one's own position in turbulent environments."[86] My contention is that the climate change migrant/refugee is a figuration of racial difference, of Blackness, that affords a similar measure of stability to the white Euro-western subject. It serves as a powerful political device which creates the impression that the undoing of racial nomos due to climate change can be governed. In this way, the figure stands as a powerful marker of white racial rule, expressing a desire to govern the climate crisis racially by bringing order to the climate change other across territory and sea.

Racial futurism also matters for the unique geography it brings into being. It constructs the relation between climate change and migration as a global problem space that sits awkwardly outside the present. In this way, it designates a kind of future threshold space between the global present and the global future. And it is here, within this awkward, virtual space, that the subject of the discourse—the climate migrant/refugee—is installed as a liminal, racial figure of this threshold. What makes the figure so unique, however, is that as a boundary object, it is situated at the limits of representation and symbolises an emergent potential. The figure of the climate migrant/refugee is not an immediately observable object. Like the figure of Blackness, it is nothingness personified. Its objectivity resides solely in its potential. It is a name given to an emergent potential dormant in the present, a potential eruption of movement or uncontainable political force akin to social death. Racial futurism is principally concerned, then, with governing this "threshold of potential,"[87] a kind of anti-Black ontopower which acts on this emergent potential prior to its actualisation. I discuss this ontopower in chapter 5. Lives and places that fall within the grasp of this mode of power are differentiated racially, however, not through the dialectical reasoning of *different from* that typically characterises the Black/white binary in the political sociology of race. Instead, they are subjectified by the racial logic of the *yet-to-come*. Lives racialised as climate migrants or climate refugees become so not because they *are* displaced by climate change but because they *might* be, even while their full actualisation as such remains impossible. Racial futurism is, thus, an anticipatory form of racial rule, a kind of "futurology" of race, to borrow a term from David Theo Goldberg.[88] It does not police the racial other so much as it strives to anticipate who will be displaced or even immobilised by climate change. Racial futurism is, in this sense, an anticipatory form of anti-Blackness, preoccupied with knowing in advance who will enter into racial categorisation, who will become a climate migration/refugee and thus who might become a problem.

Thus, we should think of racial futurism as a mode of racial rule distinct from what Goldberg calls *racial naturalism* and *racial historicism*.[89] Racial naturalism is a form of racial rule in which the racial other's apparent inferiority is said to be fixed in nature. It imagines the racial other as prehistoric and, therefore, permanently outside modern historical time. As such, because it is imagined as immutable, it is also said to be irredeemable and, therefore, permanently available to racial naturalism's governing logic of coercion. Plantation slavery and South African Apartheid are examples of this naturalist logic. Racial historicism, by contrast, treats white supremacy as an historical achievement, at the same time conceptualising the racial other as "developmentally immature." Racial historicism imagines the racial other as redeemable and, thus, available for religious conversion or, in its secular version, historical transformation. Improvement, development, education and tutelage, thus, comprise racial historicism's governing logics. Canada's residential school system is a manifestation of racial historicism, as is nineteenth-century British liberalism and much of contemporary international development. Racial naturalism and racial historicism have in common the shared fantasy that the racial other lives in some remote, anterior time. Racial futurism, by contrast, is a modality of racial rule in which the racial other is imagined to live in the future.

Finally, racial futurism is indifferent to whether the figure of the climate migrant/refugee is constructed as an object of fear, pity or welcome. This is the point about anti-Blackness. It conceptualises Blackness as completely malleable. Racial futurism assumes the unfettered availability of the figure for white renewal, regardless of whether the figure is constructed in favourable or unfavourable terms. To paraphrase Sara Ahmed, it is not a question of whether the figure of the climate migrant/refugee is conceptualised as either good or bad. What matters is how it works to secure the white human.[90] This means that the figure can be made available to all manner of racial projects. Violence, improvement, pity, welcome and charity can *all* be foisted on the figure in equal measure as white humanism strives to overcome the ontological crisis that climate change poses to humanism. Racial futurism should therefore not be mistaken as a project of racial denigration or hatred but one present in *all* formations of whiteness. Racial futurism is present, for example, in eco-fascist fantasies of "white replacement," just as it can be found in seemingly more benign formations of liberal whiteness which approach the racial other as an object to be saved, nurtured, tolerated, assimilated or integrated. Racial futurism is a form of power that links these projects in a shared commitment to the fantasy of white racial destiny, the continuation of white racial rule.

## NOTES

1. Francois Gemenne, "How They Became the Human Face of Climate Change. Research and Policy Interactions in the Birth of the 'Environmental Migration' Concept," in *Migration and Climate Change*, ed. Étienne Piguet, Antoine Pécoud, and Paul de Guchteneire (Cambridge: UNESCO Publishing and Cambridge University Press, 2011); Kumari Rigaud et al., *Groundswell: Preparing for Internal Climate Migration* (Washington, DC: The World Bank, 2018).
2. Ingrid Boas, *Climate Migration and Security: Securitisation as a Strategy in Climate Change Politics* (New York/London: Routledge, 2015); Gregory White, *Climate Change and Migration: Security and Borders in a Warming World* (Oxford: Oxford University Press, 2011); Simon Dalby, *Security and Environmental Change* (Cambridge: Polity, 2009); Michael Werz and Laura Conley, *Climate Change, Migration and Conflict: Addressing Complex Crisis Scenarios in the Twenty-First Century* (Washington, DC: Center for American Progress, 2012); Rafael Reuveny, "Climate Change-Induced Migration and Violent Conflict," *Political Geography* 26, no. 6 (2007); François Gemenne et al., "Climate and Security: Evidence, Emerging Risks, and a New Agenda," *Climatic Change* 123, no. 1 (2014); Jon Barnett, "Security and Climate Change," *Global Environmental Change* 13, no. 1. (2003); Kurt M. Campbell et al., *The Age of Consequences: The Foreign Policy and National Security Implications of Global Climate Change* (Washington, DC: Center for Strategic and International Studies and Center for a New American Security, 2007); CNA, *National Security and the Threat of Climate Change* (Alexandria: CNA Corporation, 2007); Essam El-Hinnawi, *Environmental Refugees* (Nairobi: United Nations Environment Programme, 1985); Frank Biermann and Ingrid Boas, "Preparing for a Warmer World: Towards a Global Governance System to Protect Climate Refugees," *Global Environmental Politics* 10, no. 1 (2010).
3. Vikram Odedra Kolmannskog, *Future Floods of Refugees* (Oslo: Norwegian Refugee Council, 2008); Christian Aid, *Human Tide: The Real Migration Crisis* (London: Christian Aid, 2007).
4. Romain Felli, "Managing Climate Insecurity by Ensuring Continuous Capital Accumulation: 'Climate Refugees' and 'Climate Migrants,'" *New Political Economy*

18, no. 1 (2012); Giovanni Bettini, "Climate Migration as an Adaption Strategy: De-securitizing Climate-Induced Migration or Making the Unruly Governable?" *Critical Studies on Security*, no. 2 (2014).

5. Following *Chicago* style, Black and Blackness are capitalised throughout the text. This reflects the view that Black is widely considered to be a racial and ethnic identity with a shared history, not unlike Asian, Indigenous, Latino and African. Blackness is capitalised for the same reason. White and whiteness appear throughout in lowercase as I define both as forms of power more so than forms of identity. Brown is also in lowercase, as it does not refer to a specific racial or ethnic identity, but is, instead, an umbrella term often used to group together people of colour not considered to be Black.

6. Olivia Dun and Francois Gemenne, "Defining 'Environmental Migration,'" *Forced Migration Review* 31 (October 2008): 10–11; Gregory White, *Climate Change and Migration: Security and Borders in a Warming World* (Oxford: Oxford University Press, 2011).

7. Ibid., 13–30.

8. Neel Ahuja, *Planetary Specters: Race, Migration, and Climate Change in the Twenty-First Century* (Chapel Hill: University of North Carolina Press, 2021).

9. Ellie Mae O'Hagen, "Mass Migration Is No 'Crisis': It's the New Normal as the Climate Changes," *The Guardian*, 18 August 2015; Michael Isaac Stein, "How to Save a Town from Rising Waters," *Wired Magazine*, 25 January 2018; Abrahm Lustgarten, "The Great Climate Migration Has Begun," *The New York Times Magazine*, 23 July 2020; "How Climate Change Will Reshape America," *The New York Times Magazine*, 15 September 2020; and "How Russia Wins the Climate Crisis," *The New York Times Magazine*, 16 December 2020; Rachel Gutman, "The Three Most Chilling Conclusions from the Climate Report," *The Atlantic*, 26 November 2018; Laura Parker, "143 Million People May Soon Become Climate Migrants," *National Geographic*, 19 March 2018; Parag Khanna, "Migration Will Soon Be the Biggest Challenge of Our Time," *Financial Times*, 3 October 2021.

10. *The Edge* was produced by Emma Cameron and devised and directed by Douglas Rintoul. It was first performed at New Wolsey Theatre, Ipswich, on 8 October 2015.

11. Dipesh Chakrabarty, "Postcolonial Studies and the Challenge of Climate Change," *New Literary Review* 43, no. 1 (2012).

12. Ibid., 13.

13. Sylvia Wynter, "Unsettling the Coloniality of Being/Power/Truth/Freedom: Towards the Human, after Man, Its Overrepresentation—an Argument," *CR: The New Centennial Review* 3, no. 3 (2003); Frantz Fanon, *The Wretched of the Earth* (London: Penguin, 2001); Kay Anderson, *Race and the Crisis of Humanism* (London: Routledge, 2005).

14. Fanon, *The Wretched*, 22.

15. John Hobson, *The Eurocentric Conception of World Politics: Western International Theory, 1760–2010* (Cambridge: Cambridge University Press, 2012).

16. See, for example, Anderson, *Race*; Donna Haraway, *Symians, Cyborgs and Women: The Reinvention of Nature* (London: Routledge 1991); Edward Said, *Culture*

*and Imperialism* (New York: Vintage Books, 1994); Edward Said, *Orientalism* (New York: Vintage Books, 1979).

17. W. E. B. Du Bois, *The Souls of Black Folk: Essays and Sketches* (London: Archibald Constable & Co., 1905), 13.

18. John R. Wennersten and Denise Robbins, *Rising Tides: Climate Refugees in the Twenty-First Century* (Bloomington: Indiana University Press, 2017).

19. Ibid., 47.

20. Ibid., 10.

21. Ibid., 250.

22. Collectif Argos, *Climate Refugees* (Singapore: Tien Wah Press, 2010). For an excellent critique of Climate Refugees, see Yates McKee, "On Climate Refugees: Biopolitics, Aesthetics, and Critical Climate Change," *Qui Parle* 19, no. 2 (Spring/Summer 2011): 309–325.

23. Stuart Hall, "Race, Articulation and Societies Structured in Dominance," in *Sociological Theories: Race and Colonialism* (Paris: UNESCO, 1980), 342.

24. On the idea of collateral concepts, see Noel Castree, *Nature* (London: Routledge, 2005).

25. For an example of work that reads the historical back into "climate change," see Ian Baucom's *History 4° Celsius: Search for a Method in the Age of the Anthropocene* (Durham, NC, and London: Duke University Press, 2020); and Ahuja, *Planetary Specters*.

26. Geoff Mann and Joel Wainwright, *Climate Leviathan: A Political Theory of Our Planetary Future* (London: Verso, 2018).

27. Ibid., 79.

28. Ibid., 80.

29. Ibid., 76.

30. Ibid.

31. See, for example, Baucom, *History 4° Celsius*; Kathryn Yusoff, *A Billion Black Anthropocenes or None* (Minneapolis: University of Minnesota Press, 2018); Ghassan Hage, *Is Racism an Environmental Threat?* (Cambridge: Polity Press, 2017); Yasmin Gunaratnam and Nigel Clark, "Pre-Race, Post-Race: Climate Change and Planetary Humanism," *Darkmatter Journal* 9, no. 1 (2012): n.p.; Nancy Tuana, "Climate Apartheid: The Forgetting of Race in the Anthropocene," *Critical Philosophy of Race* 7, no. 1 (2019): 1–31; Axelle Karera, "Blackness and the Pitfalls of Anthropocene Ethics," *Critical Philosophy of Race* 7, no. 1 (2019): 32–56.

32. Giovanni Bettini, "Climate Barbarians at the Gate? A Critique of Apocalyptic Narratives on 'Climate Refugees,'" *Geoforum* 45 (2013).

33. Ibid., 65.

34. I take this concept of "operating system" from Jarius Victor Grove, *Savage Ecology: War and Geopolitics at the End of the World* (London: Duke University Press, 2019).

35. IPCC, *Policymaker Summary of Working Group II (Potential Impacts of Climate Change)* (Bonn: Intergovernmental Panel on Climate Change, 1990).

36. El-Hinnawi, *Environmental Refugees*.

37. James Morrissey, *Environmental Change and Forced Migration: A State of the Art Review.* Discussion paper (Oxford: Refugee Studies Centre, Oxford University, 2009); Richard Black, "Environmental Refugees: Myth or Reality?," in *New Issues in Refugee Research* (Geneva: UNHCR, 2001).

38. Patricia Saunders, "Environmental Refugees: The Origins of a Construct," in *Political Ecology: Science, Myth and Power*, ed. P. Stott and S. Sullivan (London: Arnold Publishers, 2000), 218–46.

39. Peter Schwartz and Doug Randall, *An Abrupt Climate Change Scenario and Its Implications for United States National Security* (Emeryville, CA: Global Business Network, 2003).

40. United Nations Security Council, *Letter Dated 5 April from the Permanent Representative of the United Kingdom of Great Britain and Northern Ireland to the United Nations Addressed to the President of the Security Council*, S/2007/186 (New York: United Nations Security Council, 2007).

41. United Nations Framework Convention on Climate Change, *The Cancun Adaptation Agreements*, FCCC/CP/2010/7/Add.1, Decision 1/CP.16 (Cancun: UNFCCC, 2010).

42. Jane McAdam, "From the Nansen Initiative to the Platform on Disaster Development: Shaping International Approaches to Climate Change, Disasters, and Displacement," *University of New South Wales Law Journal* 39, no. 4 (January 2016): 1518–46.

43. Sarah Nash, *Negotiating Migration in the Context of Climate Change: The Coming of Age of an Area of Policymaking* (Bristol: Bristol University Press, 2019).

44. Rigaud et al., *Groundswell*, xiv.

45. Norman Myers, "Environmental Refugees in a Globally Warmed World," *BioScience* 43, no. 11 (1993); IPCC, *Economic and Social Dimensions of Climate Change: Contribution of Working Group II to the Second Assessment Report of the IPCC*, Intergovernmental Panel on Climate Change (Cambridge: Cambridge University Press, 1995).

46. Sir Nicholas Stern, *Stern Review on the Economics of Climate Change* (London: HM Treasury, 2006).

47. Oli Brown, "The Numbers Game," *Forced Migration Review* 31 (October 2008).

48. Aid, *Human Tide.*

49. Examples of this are too plentiful to list. See, for example, Black, "Environmental Refugees"; Gemenne, "How They Became the Human Face of Climate Change."

50. IPCC, *Fifth Assessment Report, Working Group II: Impacts, Adaptation, and Vulnerability* (Geneva: IPCC, 2014), 67.

51. Michael Hulme, "Reducing the Future to Climate: A Story of Climate Determinism and Reductionism," *Osiris* 26, no. 1 (2011); William B. Meyer and Dylan M. T. Guss, *Neo-Environmental Determinism: Geographical Critiques* (London: Palgrave Macmillan, 2017).

52. Callum Nicholson, "Climate Change and the Politics of Causal Reasoning: The Case of Climate Change and Migration," *The Geographical Journal* 180, no. 2 (2014).

53. Robert McLeman, *Climate and Human Migration: Past Experiences, Future Challenges* (Cambridge: Cambridge University Press, 2014); Robert McLeman and François Gemenne, "Environmental Migration Research: Evolution and Current State of the Science," in *Routledge Handbook of Environmental Migration and Displacement*, ed. Robert McLeman and François Gemenne (Abingdon: Routledge, 2018), 3–16.

54. Nicholson, "Climate Change and the Politics of Causal Reasoning," 152.

55. Colin P. Kelley et al., "Climate Change in the Fertile Crescent and Implications of the Recent Syrian Drought," *Proceedings of the National Academy of Sciences of the United States of America* 112, no. 11 (2015).

56. Jan Selby et al., "Climate Change and the Syrian Civil War Revisited," *Political Geography* 60 (2017).

57. Katherine McKittrick, "Plantation Futures," *Small Axe* 17, no. 3 (2013); Christina Sharpe, *In the Wake: on Blackness and Being* (Durham, NC: Duke University Press, 2017).

58. Mann and Wainwright, *Climate Leviathan*, 76.

59. Andrew Baldwin, "Orientalising Environmental Citizenship: Climate Change, Migration and the Potentiality of Race," *Citizenship Studies* 16, nos. 5–6 (2012): 625–640; Andrew Baldwin, "Racialisation and the Figure of the Climate Change Migrant," *Environment and Planning A* 45, no. 6 (2013): 1474–1490; Andrew Baldwin, "Premediation and White Affect: Climate Change and Migration in Critical Perspective," *Transactions of the Institute of British Geographers* 41, no. 1 (2016): 78–90.

60. Dominic Kniveton et al., *Climate Change and Migration: Improving Methodologies to Estimate Flows* (Geneva: International Organization of Migration, 2008); Robert McLeman, "Developments in Modelling Climate Change-Related Migration," *Climate Change* 117, no. 3 (2013).

61. Campbell et al., *The Age of Consequences*; U.K. Foresight, *Migration and Global Environmental Change (2011) Final Project Report* (London: The Government Office for Science, 2011); Rigaud et al., *Groundswell.*

62. White, *Climate Change and Migration*, 4, my emphasis.

63. Some readers will recognise that my use of the concept anti-Blackness is not without controversy. Anti-Blackness is, for many, an encompassing term which designates the way social or political life is structured racially. By this meaning, anti-Blackness is a structural condition, not an individual expression of hatred or prejudice. For the most part, I follow this convention throughout this book. The controversy lies, however, insofar as I have opted to use Frank Wilderson's conception of anti-Blackness to *explain* anti-Blackness. And this matters because while his explanation is clear and, to me, useful for explaining racial futurism, Wilderson's usage has been the subject of rather intense contestation along with the theory of Afropessimism of which it forms an important part. Examining these controversies in full lies well beyond the scope of this text. I do, however, attempt to clarify throughout my use of anti-Blackness and its relationship to race and racism. Suffice to say, I am drawn to Wilderson's conceptual description of anti-Blackness as a form of ontological violence. I hesitate, however, to endorse his "pessimistic" view on the intractability of anti-Blackness on the simple grounds that to do so would seem to preclude the

possibility of its structural reversal. That said, however, if there is political power to be found in Wilderson's conception of anti-Blackness, this, for me, rests in the way anti-Blackness can be understood to operate on and through the unconscious. See Frank Wilderson III, *Afropessimism* (New York: Liveright, 2020), and *Red, White, and Black: Cinema and the Structure of US Antagonism* (Durham, NC, and London: Duke University Press, 2010). For critiques of *Afropessimism*, see, for example, Jesse McCarthy, "On Afropessimism," *Los Angeles Review of Books*, 20 July 2020; Gloria Wekker, "Afropessimism," *European Journal of Women's Studies* 28, no. 1 (2020).

64. Ruthie Gilmore, *Golden Gulag: Prisons, Surplus, Crisis, and Opposition in Globalizing California* (Berkeley: University of California Press), 28.

65. Kathryn Yusoff offers an excellent starting point for tracing the linkages between these phenomena in *A Billion Black Anthropocenes or None*.

66. Calvin L. Warren, *Ontological Terror: Blackness, Nihilism, and Emancipation* (Durham, NC: Duke University Press, 2018).

67. For clarification, following Wilderson, I use anti-Blackness to describe a form of ontological violence, in which the figure of the Human and the Slave/Black comprise a structural antagonism. In contrast, following Stuart Hall, I understand "race" to be a category that naturalises the structural condition of anti-Blackness.

68. I borrow the notion that the political adapts to climate change from Mann and Wainwright, *Climate Leviathan*.

69. IPCC, *Fifth Assessment Report, Working Group II*, 771; Jane McAdam, "Refusing 'Refuge' in the Pacific," in *Migration and Climate Change*, ed. Étienne Piguet, Antoine Pécoud and Paul Gucheteneire (Cambridge: Cambridge University Press, 2011), 102, as cited in ibid. 771.

70. Leon Sealey-Huggins, "Between Apocalypse and Survival: The Violence of Climate Breakdown in, and for, the Caribbean" (Keynote Lecture: *Political Ecology of the Far-Rightt*, Lund University, 15–17 November 2019).

71. Alexander Anievas, Nivi Manachanda and Robbie Shilliam, eds., *Race and Racism in International Relations* (London: Routledge, 2015); Geeta Chowdhry and Sheila Nair, eds., *Power, Postcolonialism and International Relations* (London: Routledge, 2004); Hobson, *The Eurocentric Conception of World Politics*; Marilyn Lake and Henry Reynolds, *Drawing the Global Colour Line: White Men's Countries and the International Challenge of Racial Equality* (Cambridge: Cambridge University Press, 2008); Sayjay Seth, ed., *Postcolonial Theory and International Relations* (Abingdon: Routledge, 2013); Robert Vitalis, *White World Order, Black Power Politics: The Birth of American International Relations* (Ithaca, NY: Cornell University Press, 2015); Bill Schwarz, *Memories of Empire, Volume 1, the White Man's World* (Oxford: Oxford University Press, 2011).

72. Gemenne, "How They Became the Human Face of Climate Change"; Rigaud et al., *Groundswell.*

73. Brian Massumi, "National Enterprise Emergency: Steps towards an Ecology of Powers," *Theory, Culture, and Society* 26, no.1 (2009).

74. Ibid., 164–67.

75. Stephen Collier, "Topologies of Power: Foucault's Analysis of Political Government beyond 'Governmentality,'" *Theory, Culture, and Society* 26, no. 6 (2009).

76. Thomas Hobbes, *Leviathan* (London: Penguin Classics, 1651 [1968]), 185. This phrase sometimes appears as "war of all against all."

77. Paul Gilroy, *Darker than Blue: On the Moral Economies of Black Atlantic Culture* (Cambridge, MA: Harvard University Press, 2011).

78. Carl Schmitt, *The Nomos of the Earth*, trans. G. L. Ulmen (New York: Telos Press, 2003 [1950]).

79. Melinda Cooper, "Insecure Times, Tough Decisions: The Nomos of Neoliberalism," *Alternatives* 29 (2004).

80. I borrow the concept "pre-sentiment of disorder" from Schwarz, *Memories of Empire, Volume 1*.

81. Kathryn Yusoff and Jennifer Gabrys, "Climate Change and the Imagination," *Wiley Interdisciplinary Reviews: Climate Change* 2 (July/August 2011); Ben Dibley and Brett Neilson, "Climate Crisis and the Actuarial Imaginary: 'The War on Global Warming,'" *New Formations* 69 (2010); Timothy W. Luke, "The Climate Change Imaginary," *Current Sociology* 63, no. 2 (2015).

82. This is not to suggest that the concept of sovereignty does not matter for the exercise of racial futurism. I am just not convinced it matters in the way commentators often suggest when they say, for example, that states are responding to climate change by militarising border security. On this point, see, for example, Todd Miller, Nick Buxton and Mark Akkerman, *Global Climate Wall: How the World's Wealthiest Nations Prioritize Borders over Climate Action* (Amsterdam: Transnational Institute, 2021). It is also the case that militarised border security long pre-dates any discussion about climate change. Nor is climate change ever used explicitly to justify militarised border security or walling. Rather, I concur with Wendy Brown, *Walled States, Waning Sovereignty* (Brooklyn, NY: Zone Books, 2010), on this point that walling is symptomatic of waning political sovereignty resulting from the increased circulation of transnational forces (e.g., ideas, contraband, migrants, money and religious fealty) that are the direct consequence of globalisation.

83. Anderson, *Race*; Lake and Reynolds, *Drawing the Global Colour Line*; Anne McClintock, *Imperial Leather: Race, Gender and Sexuality in the Colonial Contest* (New York: Routledge, 1995); Schwarz, *Memories of Empire, Volume 1*.

84. McClintock, *Imperial Leather*, 26.

85. McKittrick, "Plantation Futures," 6.

86. Paul Gilroy, *Postcolonial Melancholia* (New York: Columbia University Press, 2005), 6.

87. Brian Massumi, *Ontopower: War, Powers, the State of Perception* (Durham, NC, and London: Duke University Press, 2015), 3.

88. David Theo Goldberg, *The Threat of Race: Reflections on Racial Neoliberalism* (Oxford: Wiley-Blackwell, 2009), 365.

89. David Theo Goldberg, "Racial Rule," in *Relocating Postcolonialism: A Critical Reader*, ed. David Theo Goldberg and Atto Quayson (Oxford: Blackwell Publishers, 2002).

90. Sara Ahmed, *Strange Encounters: Embodied Others in Post-Coloniality* (London: Routledge, 2000).

## *Chapter 2*

# The Racial Other of Climate Change

The racialisation of the figure of the climate migrant/refugee is a central attribute of racial futurism. What does it mean to say the figure is racialised? To answer this question, consider the popular film *Climate Refugees*.[1] Midway through the film a pattern of arrows is superimposed onto a spherical image of Earth. Originating in Africa, Mexico, the Middle East and Oceania, and converging on Europe and North America, the arrows are a crude attempt to map future migration in the context of climate change. Not unlike Wennersten and Robbins's racism discussed in the opening chapter, the arrows are a clear attempt by the filmmaker to convey the impression that migration in the context of climate change will involve the large-scale movement of people of colour from the Global South to the wealthy countries of the North. In this way, the arrows not only exaggerate the relationship between climate change and migration, giving visual form to an improbable future, but they also exploit a long-standing fear of immigration that has haunted Western political culture from at least the late nineteenth century. It might, therefore, be tempting to conclude that the figure of the climate migrant/refugee is racialised precisely because of representations such as this, images that construct the figure as an object of fear or alarm. And yet while this conclusion would not be incorrect, it is only partially accurate because it equates racialisation with a very narrow interpretation of racism. It wrongly assumes that racialisation is a form of hostility directed at those imagined to be racially different, in this case migrants, as opposed to a system of power which serves the purpose of separating out those marked for domination. Audrey Kobayashi and Linda Peake describe racialisation as the "material processes and the ideological consequences of the construction of 'race' as a means of differentiating, and valuing, 'white' people above those of color."[2] To say, then, that the figure of the climate migrant/refugee is racialised is not simply to say that the figure is vested with racial meaning but that the figure serves the important purpose of defining the boundaries of white Euro-western subjecthood in the context of climate change.

This chapter examines not only the ways in which the figure of the climate migrant/refugee is inscribed with racial meaning but also how such racial inscriptions value whiteness over and above Blackness. The bulk of this chapter examines a series of related racial tropes which perform this work: *naturalisation*, *the loss of political status*, *recognition*, *excess* and *ambiguity*. All of these, I argue, constitute the onto-epistemological structure of whiteness that organises the climate change and migration relation. Following Alexander Weheliye, I approach whiteness not as an identity so much as "a series of hierarchical power structures that apportion and delimit which members of the *Homo sapiens* species can lay claim to full human status."[3] The chapter analyses how each trope does the political work of sustaining this hierarchical power structure amidst the global environmental crisis. It does this by examining how each delimits the figure of the climate migrant/refugee and, thus, how each works to expel the figure from the category of the human.[4] Ultimately, through these tropes, the figure is racialised *as* Black.

The chapter draws from David Theo Goldberg's observations that a series of "preconceptual elements of racial discourse" undergird racist expression.[5] Such preconceptual elements are the epistemological glue that prevents the structural condition of whiteness from unravelling. They include assumptions about "classification, order, value, and hierarchy; differentiation and identity, discrimination and identification; exclusion, domination, subjection and subjugation; as well as entitlement and restriction."[6] Racist expression is always accompanied by assumptions that combine these preconceptual elements in some way. They are important to us here because they can help to clarify how assumptions about race are central to the forms of knowledge that organise climate migration discourse. These preconceptual elements are not, however, pregiven or natural but "manifestations of power relations vested in and between historically located subjects, and they are effects of a determinate social history."[7] Part of the task of this chapter is to consider how these tropes—naturalisation, the loss of political status, recognition, excess and ambiguity—are themselves vested in, and the effects of, specific social histories of racial domination.

To understand how the figure of the climate migrant/refugee acquires racial meaning, therefore, requires that we situate these tropes within what Sherene Razack calls "a history of the encounter between the West and its racial others."[8] In particular, the chapter shows how these tropes have their origins in the racial violence of European colonialism, transatlantic slavery or both. Understanding the historical origins of these tropes is especially important because without this history the risk is that the figure of the climate migrant/refugee is objectified. Indeed, one of the very hallmarks of whiteness is the way it seeks to secure itself by forgetting the historical racisms that have been so essential to the way it operates as a system of power. Recovering at least some of this history by historicising these tropes will allow us to challenge the way in which the white humanism sustained by the discourse on climate change and migration is imagined as internal to itself. Instead, by thinking through the ways these tropes are tethered to genealogies of racist violence, we stand to make the discourse on climate change and migration and, thus, the wider discourse of climate change more accountable to those histories.

## NATURALISATION

Nature is arguably the most important category by and through which the figure of the climate migrant/refugee is racialised. This occurs when the figure is naturalised, that is, when the migrant's historical substance is either subordinated or erased. As Thomas Nail argues, "It is not a natural fact that the history of migrants has rendered the migrant ahistorical, as Hegel argues—it is the violence of states that has rendered the migrant ahistorical."[9] The idea that the migrant's ahistoricism is political rather than natural parallels a long-standing claim in the humanities and critical social sciences that race and nature are twinned in the exercise of power.[10] Achille Mbembe puts this in no uncertain terms when he writes that "what makes the savages different from other human beings is less the color of their skin than the fear that they behave like a part of nature, that they treat nature as their undisputed master."[11]

Insofar as humanism rests on the preconceptual distinction between nature and society, it should hardly be surprising that "naturalised" life gets pushed to the margins of humanist thought.[12] This, after all, is the fundamental procedure of racial dehumanisation, where claims to humanism rest on casting the naturalised other outside modernity.[13] Thus, for Mbembe, people are marked out for racial violence not because of their skin colour but because they are thought to behave like a part of nature, the true marker of 'race.' These insights are profoundly relevant for any attempt to come to terms with claims that something called the 'environment' is a driving force in human migration. As proxies for nature, environment and climate are the key mechanisms which render the migrant ahistorical and, thus, racial. Indeed, as Paul Gilroy writes, racism "rests on the ability to contain Blacks in the present, to repress and to deny the past."[14]

To consider how the figure of the climate migrant/refugee is naturalised, let us start with the widespread use of water metaphors within the discourse. Water is a long-standing trope in Western political discourse, signifying the uncontainable, irrational and inassimilable.[15] Floods and tides, even tsunamis, are all commonly used metaphors to describe the figure of the climate migrant. Examples of water metaphors are widespread throughout the literature on climate change and migration and are simply too numerous to list here in full,[16] although one recurring motif is that of the "rising tide," noteworthy examples of which include John Wennersten and Denise Robbins's *Rising Tides: Climate Refugees in the Twenty-First Century* and chapter 6 in Lester Brown's *World on the Edge* titled "Environmental Refugees: The Rising Tide." It remains an open question whether the authors of these texts are aware of Lothrop Stoddard's *The Rising Tide of Color against White World-Supremacy*. First published in North America and Europe in 1920, it is a classic in the genre of racist geopolitics.[17]

How do water metaphors racialise? One of the founding logics of race thinking is that *Homo sapiens* can be subdivided into distinct groups, or "races," which are then spatialised. The practice of fixing bodies to specific territories is the precondition for a racialised social order in which every*body* is imagined to have its own distinctive place in the "family of man" as given by nature. Such is the underlying referent of European racial nomos, as well as a core precept of environmental determinism, the main precursor to contemporary debates about climate change and migration and a topic addressed in chapter 4. Tides, flows and floods are all powerful racial metaphors because they stand in for racial disorder by threatening to dissolve racial boundaries.[18]

But the figure of the climate migrant/refugee acquires racial meaning through more than just naturalising metaphors. As Paul Gilroy explains, "The oscillation between Black as problem and Black as victim has become, today, the principle mechanism through which 'race' is pushed outside of history and into the realm of natural, inevitable events."[19] What Gilroy means is that the racial other is conceived in the language of threat and victimhood and that its doubled status as such is naturalised. We can examine how the figure of the climate migrant/refugee is naturalised as *both* threat *and* victim by turning again to the documentary film *Climate Refugees*.[20]

Released in 2010, *Climate Refugees* sought to raise awareness about the migratory effects of climate change. It premiered at the Sundance Film Festival in 2010 and screened at numerous U.S. college campuses and film festivals worldwide and at the UNFCCC Conference of the Parties in Copenhagen in 2009.[21] It is told from the point of view of a white American filmmaker, Michael Nash, who narrates the film from the position of passionate witness, flying around the globe to document human suffering brought about by climate change and extreme weather events. At first blush, *Climate Refugees* seems laudable. After all, few things are more virtuous than raising awareness about the plight of those less fortunate. However, it is precisely the film's white saviourism that should alert us to its anti-Blackness. The authoritative voices in the film, a mix of celebrities, politicians and international legal and policy experts, all calling attention to the pending crisis of climate refugees, are almost all white Americans or Europeans, whereas the victims of climate change depicted in the film are almost all from places associated with the worst effects of climate change, such as Bangladesh, China and Tuvalu. That most of these victims are people of colour is self-evident. The point here, however, is not just that *Climate Refugees* follows a crude racial logic but that the film naturalises the migrant into an irresolvable tension between the contradictory positions of threat and victim. It is not that the climate migrant/refugee is either one or the other, but that it is *always* both.

The migrant-as-threat narrative is clearly evident in the red arrows discussed earlier; future disorder bearing down on the West. It is also evident in repeated expert claims throughout the film that climate change is a source of political instability. For example, when the then U.S. senator John Kerry claims that climate change is on "a complete collision course right now with civilisation as we know it," he is referring to what he imagines will be the destabilising effect of so-called climate refugees. The migrant-as-victim narrative is also evident in the film when constructed geopolitically as an object of humanitarian concern. This is captured succinctly in an anecdote told by Nash himself about a boy he met while in Bangladesh. In Nash's telling, and here the film audience is obliged to take Nash's testimony on faith for the boy remains voiceless in the film, the boy approaches Nash to ask him a favour: to tell "everybody [back in America] that Bangladesh is not going to survive unless America helps them." Here, Bangladesh is constructed as the wilful recipient of American benevolence; America's paternalistic foreign policy conjured through the infantilisation of Bangladesh in the image of the mute boy. And yet whatever else might be said of these two narratives, the tension between them remains unresolved throughout the film. What matters is that the figure is *both* the menacing red arrow (threat) *and* the mute boy (victim).

How is this split naturalised into the film's narrative? In *Climate Refugees*, the split subject of the climate refugee is naturalised when its displacement is said to be a function of *external* nature, for example, rising sea levels or extreme weather, as opposed to the migrant's own *internal* decision-making. In this case, the climate refugee is said to be a *forced* migrant where the force displacing the migrant is the biophysical world itself. Here, the migrant becomes a migrant not by rational choice but because he is made to "behave like a part of nature," because he is subsumed by nature as though it were his undisputed master. To be spared a life of violence or despair wrought by external nature, the mute boy's only hope is not his immediate social network, nor the Bangladeshi state. Instead, the boy must call upon the agents of history (American foreign policy), for they alone possess the capacity to seize control of the situation. The boy's social positioning, *his* history, is of no consequence when external nature is imagined as his undisputed master. He loses everything, even his right to speak.

The racial doubling of the climate migrant/refugee is further naturalised when the migrant's history is concealed. In the film, for example, Nancy Pelosi (then Speaker of the U.S. House of Representatives) and Lester Brown attribute Hurricane Katrina to climate change, which they then use to explain why so many New Orleans residents were displaced by the hurricane. But surely the history of racism in New Orleans matters greatly for how we should interpret the effects of Hurricane Katrina.[22] In a move typical of white historical denial, *Climate Refugees* neglects this all-important history. To paraphrase Nail, it is not a natural fact that climate migrants are rendered ahistorical; it is the violence of states, in this case, personified by Pelosi and Kerry.

An important consequence arises from the foregoing. If naturalisation is important in the exercise of racial power, this is because naturalisation conceals the power relations that set the European apart from its racial other. In this way, the power of naturalisation lies in its capacity to remain invisible as a form of power.[23] Naturalisation is, thus, central to anti-Blackness insofar as it places an ahistorical distance between white subject and Black object. Both are made to appear self-evident such that whiteness is seen to bear no relation to Blackness. What this means for climate migration discourse is that the threat/victim posture of the figure of the climate migrant/refugee has the appearance of being explainable solely in relation to itself. It appears to be a natural fact and not an artefact of power.

## LOSS OF POLITICAL STATUS

The loss of political status is another powerful trope through which the figure of the climate migrant/refugee is racialised. This is because the loss of political status is a defining attribute of Blackness. Achille Mbembe (2003) puts this most succinctly when he writes that "the slave condition results from a triple loss: loss of a 'home,' loss of rights over his or her body, and loss of political status."[24] He reminds us, first, that within the scope of race thinking full political subjecthood is an attribute that is said to belong only to those who are free. But he also reminds us that the act of suspending the other's political status is a defining attribute of racism. Here we can understand how the formation of white European humanism is intimately bound up with the act of suspending the other's political status and eventually positioning the other into the category of race. In this way, the loss of political status should be understood not simply as a racial marker. It is also an index of white racial power. When the other is said to be without political status, we are in the presence of anti-Blackness.

Reference to the loss of political status within the discourse on climate change and migration arises in at least two different scenarios. It applies, first, to low-lying island states where rising sea levels are said to threaten the disappearance of entire state territories. In this scenario, the extinction of a territorial state could leave its inhabitants effectively stateless.[25] Although it remains debatable whether such inhabitants would *actually be* stateless (they would still be members of a political community, just one whose territorial integrity is compromised), it *appears* that their political status would be called into question insofar as the territory which grounds their rights of citizenship would become uninhabitable.[26] An example of this potential loss of political status is the citizens of low-lying Pacific Island states. It is widely assumed, but not inevitable, that those living in the Pacific may need to relocate outside their state territories due to rising sea levels.[27] It is often precisely this scenario that grounds the erroneous assumption that the citizens of Pacific Island states are potential climate refugees. We should not be surprised to learn, however, that Pacific Island ambassadors to the United Nations have refused to use this label on the grounds that to accept it would be to abandon their citizens by suspending their political subjecthood.[28] The second, related, scenario arises in instances of potential transboundary forced migration, as this would presumably entail forfeiting whatever rights and protections an individual or group may have enjoyed by virtue of being a member of a territorially defined political community.

The fullest expression of the loss of political status is found in the legal discourse on climate migration, where it is readily acknowledged that a "legal gap" exists in the matrix of international law as it pertains to both scenarios.[29] As the legal scholar Maxine Burkett puts it, "These individuals and communities exist in a veritable legal no-man's land."[30] The core issue here is that no specific law or corresponding rights exist for this class of person.[31] As such, this is a class of person who may at some uncertain moment in the future exist without clearly specified rights, a specific kind of *potential* non-citizen, potentially out of place in the Westphalian order of nation states. This is not to argue that this person has no rights. Legal scholars have been trying, albeit provisionally, to identify already-existing bodies of law that could be invoked to secure the rights of those whose political status might be compromised as a result of climate change.[32] It is simply to recognise that the figure of the climate migrant/refugee appearing in both scenarios is said to exist in a state of potential in which the loss of political status *could* arise at some point in the future.

It is precisely this legal gap, moreover, that has motivated attempts to devise an international legal regime to safeguard transboundary migrants displaced by climate change.[33] The Nansen Initiative was one such effort. Set up by the governments of Switzerland and Norway to establish a protection agenda for cross-border displaced persons in the context of disasters and climate change, the Nansen Initiative was a laudable attempt on the part of the international humanitarian community to fill the gap in legal protections that disasters and climate change open up.[34] But even while such efforts may be necessary and driven by the best of intentions, it is important that we not lose sight of the fact that even this high point of international humanitarian law is bound up in racial power relations. For, at the core of this effort, not only do we find a preconceptual distinction between political order and disorder, between those whose political status does not threaten political order and those who are said to threaten that order by virtue of their absent political status. Embedded within this relation is *also* the power to distinguish between those with political status and those without. This is the power to represent entire communities of people, or for that matter entire regions of the world, outside the category of the political by displacing them into a zone of potential legal ambiguity. It is here, I argue, in the power to draw this distinction that the operation of racial power is present in the trope of the loss of political status.

We might be tempted to turn to Giorgio Agamben for insight into this form of racialisation. It would appear after all that the figure of the climate migrant/refugee bears a striking resemblance to that of *homo sacer*, which, according to Agamben, is *the* form of politically disqualified life, or bare life, at the heart of political sovereignty.[35] As Agamben writes, "The fundamental activity of sovereign power is the production of bare life as the originary political element and as the threshold of articulation between nature and culture, *zoē* and *bios*."[36] It would not be a huge leap to imagine the figure of the climate migrant/refugee playing a similar role in the recalibration of sovereignty under conditions of climate change. Nor would it be difficult to imagine how positioning the figure of the climate migrant/refugees into a "veritable legal no-man's land" parallels the suspension of the law that Agamben says lies at the heart of the sovereign state of exception. And if, as Agamben suggests, the camp is the spatial form of sovereignty, then it would not be a stretch to imagine how the camp might proliferate in the coming years as *the* technology for governing the figure of the climate migrant/refugee. In fact, there is good reason to believe that the camp already serves this purpose.

Yet even while we might plausibly pursue a theory of the climate migrant/refugee using Agamben's philosophy, it is nevertheless insufficient for theorising how the trope of the loss of political status functions racially.[37] There are numerous reasons for this. The most important of which, however, borrowing from Alexander Weheliye, is that for Agamben the figure of bare life (*homo sacer*), that absolute form of political disqualification, in fact transcends race.

A figure so thoroughly debased, and devoid of life and substance, *homo sacer* is an indivisible figure whose existence has become so unliveable that it can no longer be subdivided into gradients of life that would comprise "races." In this way *homo sacer* stands as sovereignty's absolute limit and, as such, a universal conception that comes to stand as the limit of the human. For Weheliye, however, since race is not part of this calculus, Agamben's theory of political disqualification is not altogether useful for thinking the delimitations of the political in racial terms.

But, if we turn to other forms of political theory where race *is* explicit, a very different conception of political loss emerges. In Black studies, for example, the plantation economy is considered to be foundational to the formation of humanism, the very ground on which political subjecthood is guaranteed. In reference to the transatlantic slave trade, Christina Sharpe writes, for example, that "Black people ejected from the state become the national symbol for the less-than-human being condemned to death."[38] This is also Mbembe's point—the loss of political subjecthood is a condition of racial abjection. We might also consider this insight in relation to critiques of liberalism which recognise that notions of right, freedom and subjecthood could only ever be established because of racial slavery not in spite of it. In this sense, racial slavery is not liberalism's aberration but its founding condition.[39] As Domenico Losurdo puts it, "The rise of liberalism and the spread of racial chattal slavery are the product of a twin birth."[40] Similarly, in her treatment of renaissance humanism and "Man as the Rational Self and political subject of the state," Sylvia Wynter explains how modern political subjectivity "was to be effected only on the basis of" colonial and imperial power, "one of the major empirical effects of which would be 'the rise of Europe' and its construction of the 'world civilization' on the one hand, and, on the other, African enslavement, Latin American conquest, and Asian subjugation."[41] What these brief but important examples allow us to understand is that by situating the trope of the loss of political status in the "history of the encounter between the West and its racial other,"[42] we can better appreciate how modern political subjectivity took shape historically in relation to its antithesis and how the loss of political status is a synonym for Blackness.[43]

These examples of political loss as Black abjection put into sharp relief how we might understand the racialisation of the figure of the climate migrant/refugee. For when we acknowledge that the history of suspended political status is bound up with the history of racist exploitation, we enter a tradition of thought that refuses to accept the loss of political status as naturally occurring. We acknowledge that suspending the other's political status by positioning them in a legal no man's land through reference to the "climate refugee" concept is to exercise a form of power that has long been a hallmark of white European racism. It is to acknowledge, moreover, that those who enact this form of power do so not by virtue of some inherent responsibility but because they presume it their right to do so. This is why the Pacific Island ambassador's refusal to accept the label "climate refugee" is so powerful. It reminds us that the suspension of the other's political status is *always* a conscious choice and thus an historical achievement. Even if the suspension of political status is invoked hypothetically to make a point about the coming severity of climate change, it is *still* the suspension of the other's political status. We can put the sheer audacity of this power into perspective when considered in light of estimates which predict that, by 2050, 200 million people will be forced to relocate because of climate change.[44] What kind of power is it that allows such vast numbers of people in the Global South to be constructed on the verge of losing their political status even while the vast majority are already *de jure* members of a political community? A form of racial power which originates in humanism and which lays claim to the very movement of people at a moment when the habitability of Earth is no longer a certainty.

## RECOGNITION

Recognition is a third trope for considering how the climate migrant/refugee is racially invested. However, unlike naturalisation and the loss of political status, recognition is an unlikely category for thinking race and racism. This is because recognition is often associated with the acquisition of rights and the alleviation of injustice. This is certainly the case for the discourse on climate change and migration where all manner of demands are made to positively recognise the climate migrant/refugee, not least in law. But, as Glen Coulthard argues, the politics of recognition extends well beyond the liberal acquisition of rights. In fact, for Coulthard and others, recognition is never a neutral political category, nor an impartial remedy for injustice, but is best conceived as a mechanism for political subjugation. Even while recognition may offer those who seek it some measure of freedom from oppression, recognition alone is never sufficient for reversing the very power relations that draw these forms of oppression into being.[45] In Coulthard's case, this means that Indigenous peoples' struggles for recognition in the settler colonial context of Canada will never resolve the power asymmetry constitutive of Indigenous-settler relations inasmuch as these are structured by settler colonial law, one of the core functions of which is the protection of private property through the territorial dispossession of Indigenous people. But Coulthard goes further. He argues that "instead of ushering in an era of peaceful coexistence," the concept of recognition, in his case the recognition of Indigenous sovereignty by the Canadian state, "promises to reproduce the very configurations of colonialist, racist, patriarchal state power that Indigenous peoples' demands have historically sought to transcend."[46] Thus, far from ameliorating oppression, recognition can make it worse.

The preconceptual element that draws recognition into the orbit of race is an imagined hierarchy between, on the one hand, the subject of recognition—the subject to be recognised—and, on the other, a subject authorised to recognise the other. In this sense, recognition can be understood to interpellate subjugated persons. By adopting a congenial disposition, the subjugated individual becomes recognisable to those who claim authority over them. When we turn to Frantz Fanon, Coulthard's main inspiration, the raciality at stake in the political concept of recognition becomes more obvious. For Fanon, the coloniser exercises his authority by imposing on the colonised a distorted sense of being, one that negates the lived experience of the colonised and affirms the authority of the coloniser. Colonial power for Fanon operates when colonised subjects internalise this inferior self-image and so come to misrecognise themselves through a distorted consciousness. Recognition emerges in this context as a demand that one be acknowledged *on one's own terms*, the insistence that one *not* be misrecognised. The trouble, however, and this is Fanon's point, is that recognition is granted only so long as it reaffirms hegemonic power, hence Coulthard's claim that recognition reproduces the power it promises to overturn.[47]

While it is important to acknowledge that his theorisation of recognition is specific to Canadian settler colonialism, Coulthard's insights are nevertheless pertinent to our discussion about how the figure of the climate change migrant is inscribed with racial meaning. This is so, given the privileged status recognition holds in the discourse, especially the legal discourse pertaining to notions of the "climate refugee." As discussed in the previous section, the legal discourse that surrounds the figure of the climate migrant/refugee unfolds largely as a concern that existing law, oftentimes international law, fails to recognise the unique attributes of the figure. The often-made claim is that existing refugee law is handicapped to address the plight of those displaced by climate change, since the 1951 Refugee Convention only applies in instances of persecution owing to "reasons of race, religion, nationality, membership of a particular social group, or political opinion."[48] This in turn has inspired efforts to fill this legal gap, such as the Platform for Disaster Displacement (formerly the Nansen Initiative) or the U.K.-based Environmental Justice Foundation, which has been at the forefront of efforts to have the "climate refugee" recognised in European law.

Frank Biermann and Ingrid Boas's widely cited paper "Preparing for Warmer World: Towards a Global Governance System to Protect Climate Refugees" offers another example.[49] While Biermann and Boas never elaborate on their use of the term recognition, like other similar accounts, their proposed protocol hinges on the claim that existing global governance fails to recognise "climate refugees." At first glance, the protocol has obvious appeal. Like the film *Climate Refugees*, Biermann and Boas seem positively motivated to ensure that the category "climate refugee" is legally recognised, because once enshrined in law it can be mobilised to remedy injustice. Indeed, a global protocol that recognises and protects "climate refugees" may well offer individuals some measure of relief from the oppressions that may result from climate change. But, as we know from Coulthard and Fanon, recognition is never so innocuous, as it often reproduces the power it promises to reverse. This is evident in Biermann and Boas's argument where a well-intentioned attempt to positively recognise the legal rights of those said to be most at risk of being displaced by climate change can be reinterpreted as a reaffirmation of Euro-western humanism.

The preconceptual element at stake in their argument is that of racial hierarchy. Biermann and Boas are unambiguous about those they see as the beneficiaries of their protocol. Like Wennersten and Robbins, they state flatly that their analysis is restricted "to climate refugees in Africa, Asia, Latin America, and Oceania" on the grounds that "it is people in developing countries who are most likely to be compelled to leave their homes and communities." This they attribute to "low adaptive capacities, their often vulnerable location vis-à-vis climate change events, often high population densities, pre-existing hunger and health problems, low level of per capita income, often weak governance structures, political instability and other factors."[50] What Biermann and Boas seem to be suggesting is that only those living in countries formerly colonised by European imperial powers should be recognised as "climate refugees," a geography that instantly recalls W. E. B. Du Bois's global colour line.[51] This hierarchy is additionally evident in their rationale for the protocol. "Early support for climate refugees," they assert, "might not only attenuate human suffering; it might also prevent violent conflict." Such support, they claim, is an "investment in global security in the twenty-first century," and a precondition for "orderly and organised responses." Not only, then, does the climate refugee surface in Biermann and Boas's protocol in the racial language of threat/victim (*both* suffering *and* violent conflict), but recognising climate refugees would also be an investment in global biopolitical security. Thus, even while offering the possibility of rights, recognition in the way Biermann and Boas use it does not seem to be concerned with reversing long-standing structural inequalities between Europe and its former colonies so much as upholding the very global colour line that European imperialism imposed on the world.[52] In this way their assertion of recognition merely reaffirms a structure of power in which the West is assumed to possess the right to construct the majority world as unrecognisable and, thus, in need of recognition on terms set in and by the West.

Fortunately the Eurocentrism at stake in the climate refugee concept has not escaped the notice of critics.[53] Carol Farbotko is among those who have identified how a similar cosmopolitan imaginary structures Western representations of the Pacific Islands in relation to climate change.[54] Using Tuvalu as her case, she shows how depictions of Tuvalu in the Western media as remote, small and disappearing reaffirm a much longer imperial tradition in which Tuvalu is made to occupy a position subordinate to the West. Echoing the Pacific postcolonial scholar Epeli Hau'ofa, Farbotko's concern is that, by constructing Tuvaluans as vulnerable climate refugees, such an imperial imaginary "can also operate to silence alternative [Tuvaluan] identities that emphasize more empowering qualities of resilience and resourcefulness."[55] What Farbotko offers, then, is a critique not unlike that of Fanon or Coulthard. By affirming Western epistemology over and above that of Tuvalu, the label "climate refugee," she argues, *mis*recognises Tuvaluans by negating Tuvaluans' *actual* lived experience. The problem Farbotko identifies is that in its failure to adequately recognise Tuvaluans, the label "climate refugee" does more to divest Tuvaluans of their history and identity than it offers in the form of political or legal recognition. Furthermore, what gets overlooked when Tuvaluans are labelled "climate refugees" is an appreciation for Tuvaluan political agency, recognition that Tuvaluans are perfectly equipped to represent themselves when confronting climate change. Thus, one important effect of recognition's representational structure is that it forecloses Tuvalu as a site of politics.

This critique of Eurocentrism has resulted in various attempts to recover the neglected voice of the other within the discourse on climate change and migration.[56] Farbotko and her co-author Heather Lazrus have been at the forefront of this effort, arguing that more should be done to recognise Tuvaluans, and Pacific Islanders more generally, *on their own terms*. By adopting what they call a "listening position" and "foregrounding islander perspectives," Farbotko and Lazrus are therefore able to do something Biermann and Boas could not.[57] They are able to show that Tuvaluans are not the unwitting victims of climate change they are often made out to be. Instead, they show how migration is inextricably "bound up in [Tuvaluans'] everyday practices," not exceptional to life in Tuvalu, but part of it. Enrolled in a long history of seafaring and transnational migration, "the Tuvaluan sense of history," they tell us, "is strongly connected to mobility." And this means that Tuvaluans, far from needing to be recognised as climate refugees, are, it turns out, well positioned not only to adapt to climate change but also to do so *on their own terms*.

In many respects, Farbotko and Lazrus give a refreshing account. By decentring the Eurocentrism of the climate refugee concept, they prise open a space for a more progressive form of recognition to emerge in the climate migration debate. Rather than silencing the other, Farbotko and Lazrus insist that the subaltern ought to be heard, that the agency of those on the frontline of climate change should be revered. Yet we should approach *even this* seemingly more benign account with caution. Following Coulthard, even while Tuvalu is recognised anew, Farbotko and Lazrus's account remains insufficient for reversing the structural conditions that continue to position Tuvalu and other Pacific Islands on the margins of the global fossil fuel economy. In other words, even if recognising Tuvaluans on their own terms may offer some measure of relief from the overbearing cosmopolitan imaginary that Farbotko and others warn against, when Tuvaluans *are* recognised for their embrace of migration, this is so only to the extent it reaffirms Tuvalu's subordinate position within the international order of territorial states. Migration may well be an everyday feature of life in Tuvalu. But within this narrow band of recognition there is no claim that Tuvalu's culture of mobility will allow Tuvaluans to transcend the territorial integrity of neighbouring archipelagos, such as Samoa, Fiji or the Solomon Islands, or even New Zealand. Tuvaluan migration may well be positively recognised within the wider cultural economy of climate change, but this is so only to the extent that it leaves unchallenged the primacy of territorial sovereignty which is itself an artefact of European imperialism and one of the main mechanisms used to police global labour mobility. Recognition is, in this sense, a container of Tuvaluan politics, rather than its enabler.

What we should take from the foregoing discussion, then, is that, far from emancipatory, recognition can also serve as a mechanism of political containment bound up in the genealogies of European imperial power. Recognition may not announce itself in the language of racial hatred. Nor is it a neutral term. If anything, recognition has historically been a key concept in the liberal management of racial difference.[58] Its power lies in the capacity to place limits around difference and, thereby, establish the terms of *acceptable* difference. Recognition, thus, renders the intolerable tolerable. If part of our task is to understand how the figure of climate migration is invested with racial meaning, the politics of recognition that we find present in the wider discourse should give us pause. Recognition has an established history as a tool for political subordination, often used as a means to neutralise the political claims put forth by those said to be racially other. The demand to recognise the climate refugee in law can thus be reframed as simultaneously the power to render the other unrecognisable by situating it outside the law, while at the same time promising the opposite.

## EXCESS

In chapter 1, we considered the irresolvable epistemological disjuncture haunting the discourse on climate change and migration. There I argued that no amount of knowledge that explains migration as a function of climate change will ever overcome the fact that migration is irreducible to climate or the environment; migration is only ever multicausal. Thus, the climate-migration relation is defined, paradoxically, as the very absence of a causal relation between climate change and migration. Not so much a crisis of representation as a crisis of explanation, or even better a crisis that *exceeds* explanation. According to Wendy Brown, such is "the foundational meaning of the very term krisis," an emergency "unavailable to extant categories of thought, analysis, and politics."[59] We have already seen how the figure of the climate migrant/refugee when conceived as the excess of political categorisation is racially marked. Here we consider two additional categories of excess with which the figure is positioned outside that of the human: indeterminacy and undecidability.

Throughout the discourse on climate change and migration, there is a pervasive concern with estimating the number of climate migrants/refugees the world can expect to see in the future. Perhaps the most widely cited numerical estimate belongs to Norman Myers, who estimated that, by 2050, 200 million people will migrate due to climate change. Myers's estimate, for instance, would go on to be cited in the influential 2006 *Stern Review* and throughout the international policy literature. Other predictions include Christian Aid's estimate that 250 million people will be displaced by climate change by 2050, while Robert Nicholls estimates 50 to 200 million by 2080. A recent estimate by the World Bank claims that, by 2050, 143 million could be internally displaced due to climate change.[60]

Taken together, what all these predictions signify is the widespread disagreement about the extent to which future migration can be attributed to climate change. Discrepancies like these are to be expected, of course, given that predicting the future is inherently fraught. There is nothing unique about climate migration in this respect. What makes it unique, however, is that there is no agreed definition as to who or what qualifies as a climate migrant/refugee. Nor is there a common methodology for calculating the number of people who will be displaced by climate change. It is also true that datasets on migration are also notoriously incomplete as is the census data from countries in the Global South, which Biermann and Boas have argued will be the main sites where climate migration is set to occur.[61] However, stepping back from these empirical limitations, the point I take from these discrepancies is that the figure of the climate migrant/refugee defies easy calculation.

Its quantitative indeterminacy is emphasised across the scholarly literature on climate change and migration discourse.[62] As Biermann and Boas put it, "Even though the exact number of climate refugees is hardly certain given the diverse methodological problems . . . large migration flows over the course of this century are plausible."[63] And yet what is so intriguing about such attempts at quantitative prediction is that they produce the figure as simultaneously indisputable *and* empirically elusive. Not only then is the figure the excess of *political* categorisation; it is also the excess of *quantitative* calculation, a figure defined by its very indeterminacy.

This excessiveness is even more apparent when we consider the figure's undecidability. As we have already seen, because migration is multicausal, climate change can never be the main driver of migration.[64] The argument here is that other power geometries, such as labour markets, land tenure arrangements, ethnic and racial marginalisation and war, all complicate the migration calculus such that it becomes impossible to isolate the cause of migration in any singular sense. This critique has been in circulation for almost two decades and is now quite standard across much of the climate change and migration discourse.[65] For example, Jane McAdam and Ben Saul write that "it is inherently fraught to speak of climate change as the 'cause' of human mobility," and Richard Black, Dominic Kniveton, and Kerstin Schmidt-Verkerk. have argued that "the impacts of climate change on mobility are mediated through these different [social, political, demographic and environmental] drivers."[66] By this reasoning, distinguishing between a climate migrant/refugee and other forms of migration is impossible. A decision cannot be taken about who or what is a climate change migrant or climate refugee. What the principle of multicausality allows us to conclude, then, is that, because the cause of migration remains indeterminate, the figure of the climate change migrant/refugee is also by definition undecidable. Moreover, if decisionism is said to be one of sovereignty's defining attributes, the figure's undecidability calls even this basic formation of the political into question.

Indeterminacy and undecidability are, thus, two additional markers of difference and, as categories of excess, they too place the figure of climate migrant/refugee outside the category of the human. The normal methods of legal, scientific, statistical or demographic calculation fail to fully quantify the figure in numerical terms. So, too, the historical method fails to fully render the climate change migrant because the figure's migration is said to be primarily the result of climate change; historical explanation is demoted. What this disjuncture at the core of the discourse points towards, then, is that the figure is one of excess and, thus, a figure that evades intelligibility. Not only can the climate change migrant not be fully known, but it is also defined, paradoxically, by its unintelligibility. In this way, we can say that the figure is "objectively indeterminate": we sense its presence, but we can never actually see it.[67] It is real, but not actual, a kind of virtual excess.[68] Amitav Ghosh's description of climate change as an unthinkable phenomenon is a similar descriptor of excess.[69]

The ontological turn in Black studies can clarify how positioning the figure into a zone of unintelligible excess is synonymous with anti-Blackness and thus how the figure functions racially. Calvin Warren draws the connection most succinctly in his metaphysical discussion of the way in which the concept of nothingness can be understood to function racially. Metaphysics is the branch of philosophy concerned with explanations of being, not least of *human being*. The category of nothingness, by contrast, is that which functions as the excess, or outside, that grounds metaphysics. If metaphysics seeks to give meaning to human being, then nothingness designates the absence of meaning, the abyss of meaninglessness, which by virtue of its abyssal relation to being induces what Warren describes as ontological terror. "*Nothing* terrifies metaphysics," and so metaphysics attempts to overcome the terror of nothing by dominating it, "by turning nothing into an object of knowledge, something it can dominate, analyse, calculate, and schematize."[70] In this sense, anti-Blackness involves precisely "an accretion of practices, knowledge systems and institutions designed to impose nothing onto Blackness" whereby "anti-Blackness is anti-nothing."[71] Blackness, thus, comes to signify the terror of nothingness.

A similar ontological terror is at stake in the epistemic contradiction that defines the figure of the climate migrant/refugee. In chapter 1, we saw how the figure's ungovernability is the source of this terror. The trope of excess developed in this section allows us to extend our understanding of this terror as that which arises when the figure is positioned as the excess of intelligibility and, thus, beyond the pale of being. In this sense, the figure is endowed with terror because its ungovernability arises from its lack of intelligibility in the space of nothingness. This terror, then, becomes a racial terror inasmuch as nothingness is imposed onto Blackness, when the majority world is displaced into the non-place of nothingness.

## AMBIGUITY

We saw earlier how the ambiguous either/or status of the climate migrant/refugee was naturalised. Further consideration is given here to the way ambiguity itself operates as a racialising trope. Arguments that position Bangladesh as a site of potential Islamic fundamentalism in the context of climate change offer a good example. The journalist Robert Kaplan has argued, for example, that the effects of climate change could lead to widespread resentment in Bangladesh towards the West, given its culpability in climate change. Kaplan's contention is that, "as rural Bangladeshis flee a countryside ravaged by salinity in the south and drought in the northwest, they are migrating to cities at a rate of 3 to 4 per cent a year. Swept into the vast anonymity of sprawling slum encampments, they lose their local and extended-family links, becoming more susceptible to a form of Islam with a sharper ideological edge."[72] Although one could easily take issue with Kaplan's determinism, consider the way he writes the migrant in the future conditional tense: the so-called climate migrant *could* become embroiled in religious fundamentalism. By Kaplan's reasoning, the Bangladeshi climate migrant remains a troubling presence precisely because it embodies both disruptive *and* peaceable potential. Either the Bangladeshi migrant will generate violence or it won't, but before we can know for certain which, the figure remains indeterminate. The both/and status of the climate change migrant may help to explain why this figure is so caricatured as the embodiment of insecurity or humanitarian suffering. The status of being both/and, or what the philosopher Brian Massumi refers to as the quality of being singular-generic (as in generative rather than general) is the mark of that which is yet to be fully known. The singular-generic generates multiple possibilities through its very presence and, in this regard, is inherently unpredictable.[73] As such, one can never be sure how the singular-generic will unfold; hence one can never be fully sure what consequences the singular-generic will bring to a situation. Will the singular-generic prove unmanageable once fully known? Or will it succumb to easy containment? Will it be friend? Or will it be foe?

How does ambiguity work as a racial trope? To answer this question, consider how the climate migrant/refugee bears striking resemblance to the figure of the monster, a prophetic figure steeped in racial meaning.[74] In his Collège de France 1974–1975 lectures published under the title *Abnormal*, Michel Foucault tells us that "the frame of reference of the human monster is, of course, the law." "What defines a monster," he adds, "is the fact that its existence and form is not only a violation of the laws of society but also a violation of the laws of nature":

> Although it is a breach of the law . . . the monster does not bring about a legal response from the law. It could be said that the monster's power and its capacity to create anxiety are due to the fact that it violates the law while leaving it with nothing to say. . . . When the monster violates the law by its very existence, it triggers the response of something quite different from the law itself. It provokes either violence, the will for pure and simple suppression, or medical care and pity.[75]

Thus, for Foucault, the human monster is marked by an irresolvable ambiguity. Later in the same passage, he refers to this ambiguity as the monster's "tautological intelligibility."[76] The monster's difference exceeds comprehension, thus, forcing one to look deeply into all the little internal deviations of the monster (morphology, speech, attitude, etc.) in order to explain its difference. We dissect the monster in a fit of incomprehension. The demand that we generate more empirical evidence of climate migration and, thus, of the climate migrant, is akin to the demand that we dissect this monstrous phenomenon and learn what we can of all its interval deviations. For Foucault, moreover, it is the abnormal figure, in this case, the monster, that underpins biopolitical racism, "whose function is not so much the prejudice or defence of one group against another as the detection of all those within a group who may be the carriers of a danger to it."[77] This is the figure that must be eliminated, curtailed and contained in order for the purportedly normal population to flourish.

Jane Anna Gordon and Lewis Gordon[78] provide a similar argument about the racial substance of monsters. Monsters, they argue, can be mythopoetic, nihilistic or modern, but in each case monsters are always understood to carry a warning. My contention is that the figure of the climate migrant/refugee resembles the modern monster, which Gordon and Gordon tell us, echoing Foucault, is a figure who is "naturally unnatural." Monsters are of nature but, since they fail to comply with the *order* of nature, they fall outside it. Their ambiguous presence as simultaneously natural and unnatural thus becomes a warning, but, as is typical of monsters, a warning that goes unheard. Gordon and Gordon tell us that monsters are survivors of disaster, what they call "the sign continua of the disaster." By this they mean that monsters are the lingering effects of disaster that admonish us to confront the crisis of values materialised by the disaster itself. Thus, monsters signify the symptoms of disaster. Markers of and marked by disaster, they emerge from disastrous landscapes with a warning to confront the crisis of social relations that underpin the disaster, to confront the way in which humanity comes to know itself only through the oppressions and violent repudiations of its perceived others. Gordon and Gordon remind us, however, that too often we fail to heed the warnings that monsters signify. Thus, purified of their warnings, we blame monsters for their own collective inability to respond to the crisis of values from which they are born. Black people displaced by Hurricane Katrina are thus blamed for their "monstrous" condition, even while the disaster of Hurricane Katrina reveals a collective failure to address the crisis of race in America.

Monstrous ambiguity is a trope that is also at stake in the figure of the climate change migrant.[79] Judith Butler's considerations of what counts as a grievable life in the context of the US-led invasions of Afghanistan and Iraq offer a concise synopsis of the trope of ambiguity. Arguing that the mark of a grievable life is its recognisability, Butler's contention is that grievability applies only to those who can be recognised as a life. However, in order for a life to be recognised as a life, it must also be recognisable as such. It must have the capacity to be recognised as a life. But for Butler, and this is crucial, the condition of recognisability is intelligibility, which is to say that a life is only recognisable if it is intelligible, if it can be shorn of ambiguity. This formulation has potentially far-reaching implications for how we might conceive of the climate change migrant as an object of political and legal discourse. For if we are prepared to accept that the figure of the climate migrant/refugee is defined by its unintelligibility, by its naturalisation, its absent political status, its excess and its ambiguity, then by Butler's account the figure falls short of being a life because it lacks recognisability. Its ambiguous status as both/and bars it from the status of recognition and, hence, from law. Or we might simply say that its ambiguity is precisely what expels the figure from humanity, hence its racialisation.

## NOTES

1. *Climate Refugees: The Human Face of Climate Change*, directed by Michael Nash (Los Angeles: LA Think Tank, 2010), 89 mins.

2. Kobayashi and Peake, "Racism out of Place," 393.

3. Alexander Weheliye, *Habeas Viscus: Racializing Assemblages, Biopolitics, and Black Feminist Theories of the Human* (Durham, NC and London: Duke University Press, 2016), 19.

4. My concern in this chapter is with the racial tropes that give the figure of the climate migrant/refugee meaning. All of them construct the figure outside humanism, and in this way the figure is made to occupy the very same structural position of Slave/Black within the structural antagonism that, for Wilderson, defines humanism.

5. David Theo Goldberg, *Racist Culture: Philosophy and the Politics of Meaning* (Oxford: Blackwell, 1993), 48.

6. Ibid., 49.

7. Ibid., 49.

8. Sherene Razack, *Casting Out: The Eviction of Muslims from Western Law and Politics* (Toronto: University of Toronto Press, 2008), 6.

9. Thomas Nail, *The Figure of the Migrant* (Stanford, CA: Stanford University Press, 2016), 5.

10. See, for example, Donna Haraway, *Primate Visions: Gender, Race and Nature in the World of Modern Science* (New York and London: Routledge, 1989); Audrey Kobayashi and Linda Peake, "Unnatural Discourse: 'Race' and Gender in Geography," *Gender, Place and Culture* 1, no. 2 (1994); Donald Moore, Jake Kosek and Anand Pandian, "The Cultural Politics of Race and Nature: Terrains of Power and Practice," in *Race, Nature and the Politics of Difference*, ed. Donald Moore, Jake Kosek and Anand Pandian (Durham, NC, and London: Duke University Press, 2003).

11. Achille Mbembe, "Necropolitics," *Public Culture* 15, no. 1 (2003): 24.

12. Bruno Latour, *We Have Never Been Modern* (New York: Harvester/Wheatsheaf, 1993).

13. Wynter, "Unsettling the Coloniality of Being/Power/Truth/Freedom"; Razack, *Casting Out*.

14. Paul Gilroy, *There Ain't No Black in the Union Jack* (London: Routledge, 1991), 12.

15. Klaus Theweleit, *Male Fantasies: Volume 1: Women, Floods, Bodies, Histories* (Minneapolis: University of Minnesota Press, 1987).

16. Notable examples include: Lester Brown, "The Rising Tide of Environmental Refugees," *InterPress Service News Agency*, 22 October 2009; Lester Brown, *World on the Edge: How to Prevent Environmental and Economic Collapse* (New York and London: W.W. Norton & Co., 2011); Aid, *Human Tide*; Bonnie Docherty and Tyler Giannini, "Confronting a Rising Tide: A Proposal for a Convention on Climate Change Refugees," *Harvard Environmental Law Review* 33, no. 2 (2009); Kolmannskog, *Future Floods*; Wennersten and Robbins, *Rising Tides*; Angela Williams, "Turning the Tide: Recognizing Climate Change Refugees in International Law," *Law and Policy* 30, no. 4 (2008); Gordon McGranahan, "The Rising Tide: Assessing the Risks of Climate Change and Human Settlements in Low Elevation Coastal Zones," *Environment and Urbanization* 19, no. 1 (2007); Janos Bogardi and Koko Warner, "Here Comes the Flood," *Nature Climate Change* 138, no. 11 (2008).

17. Lothrop Stoddard, *The Rising Tide of Color against White World-Supremacy (New York: Charles Scribner's Sons, 1925); Vitalis, White World Order*.

18. However, it is also true that the absence of water can contribute to the construction of racial meaning. I don't address how this might be so, but one starting point might be to consider how people are reduced to processes of desertification and desiccation, both of which are geohistorical processes that find expression in climate migration discourse.

19. Gilroy, *There Ain't No Black in the Union Jack*, 11.

20. *Climate Refugees*, dir. Nash.

21. See http://www.climaterefugees.com/. Last accessed 14 April 2021.

22. Henry Giroux, "Reading Hurricane Katrina: Race, Class, and the Biopolitics of Disposability," *College Literature* 33, no. 3 (2006); John Protevi, *Political Affect: Connecting the Social and the Somatic* (Minneapolis: University of Minnesota Press, 2008).

23. Bruce Braun, *The Intemperate Rainforest: Nature, Culture, and Power on Canada's West Coast* (Minneapolis: University of Minnesota Press, 2002).

24. Mbembe, "Necropolitics," 21.

25. Jane McAdam, "'Disappearing States,' Statelessness and the Boundaries of International Law," in *Climate Change and Displacement: Multidisciplinary Perspectives*, ed. Jane McAdam (Oxford: Hart Publishing, 2010); Maxine Burkett, "The Nation Ex-Situ: On Climate Change, Deterritorialized Nationhood and the Post-Climate Era," *Climate Law* 2 (2011); Sumudu Atapattu, "Climate Change: Disappearing States, Migration, and Challenges for International Law," *Washington Journal of Environmental Law & Policy* 4, no. 1 (2014).

26. On the concept of territorial integrity, see Stuart Elden, *Terror and Territory* (Minneapolis: University of Minnesota Press, 2009).

27. One alternative to relocation often discussed among Pacific Islanders is the physical expansion of islands through sea reclamation and infilling.

28. Karen McNamara and Chris Gibson, "'We Do Not Want to Leave Our Land': Pacific Ambassadors at the United Nations Resist the Category of 'Climate Refugees,'" *Geoforum* 40 (2009).

29. Jane McAdam, *Climate Change, Forced Migration, and International Law* (Oxford: Oxford University Press, 2012). This is claimed in an extensive body of literature. See, for example, Darlington Mwape and Volker Türk, *Report by the Co-Chairs: Gaps in the International Protection Framework and in Its Implementation* (Geneva: United Nations High Commissioner on Refugees, 2010); Koko Warner, "Global Environmental Change and Migration: Governance Challenges," *Global Environmental Change* 20, no. 3 (2010).

30. Burkett, "The Nation Ex-Situ," 350.

31. Jane McAdam and Ben Saul, "An Insecure Climate for Human Security? Climate-Induced Displacement and International Law," in *Human Security and Non-Citizens: Law, Policy and International Affairs*, ed. Alice Edwards (Cambridge: Cambridge University Press, 2009).

32. Ibid.; McAdam, *Climate Change, Forced Migration, and International Law*.

33. Benoit Mayer, "The International Legal Challenges of Climate-Induced Migration: Proposal for an International Legal Framework," *Colorado Journal of International Environmental Law and Policy* 22 (2011); Biermann and Boas, "Preparing for a Warmer World"; Docherty and Giannini, "Confronting a Rising Tide."

34. After it was completed, the Nansen Initiative became the Platform on Disaster Displacement. See McAdam, "Nansen Initiative," and Platform on Disaster Displacement, https://disasterdisplacement.org.

35. Giorgio Agamben, *Homo Sacer: Sovereign Power and Bare Life*, trans. Daniel Heller-Roazen (Stanford, CA: Stanford University Press, 1998).

36. Ibid., 195.

37. Weheliye, *Habeas Viscus*.

38. Christina Sharpe, *In the Wake: On Blackness and Being* (Durham, NC, and London: Duke University Press, 2016), 79.

39. Barbara Jeanne Fields, "Slavery, Race, and Ideology in the United States of America," *New Left Review* 181, no. 1 (1990).

40. Losurdo, *Liberalism*, 37.

41. Wynter, "Unsettling the Coloniality of Being/Power/Truth/Freedom," 263.

42. Razack, *Casting Out*, 6.

43. Wilderson, *Afropessimism*.

44. Readers should note that an important distinction should be drawn between this particular estimate, often attributed to Norman Myers, and a series of quantitative estimates recently produced by the World Bank. See Rigaud et al., *Groundswell*, and Clement et al., *Groundswell Part 2: Acting on Internal Climate Migration* (Washington, DC: The World Bank, 2021.) While Myers's is a kind of generic reference, the World Bank confines its estimates to the so-called internal climate migrants, those

who might be displaced by climate change but who won't leave their countries of origin. A full explanation as to why the World Bank opted for this measure goes beyond my argument. I would wager, however, that part of the reason they've done so has been to create the impression that "climate migration" is a domestic political concern that can be governed using existing multilateral institutional arrangements. For the bank to cultivate the relation between climate change and migration as a transnational problem would, presumably, involve a collective international response, which would be much more challenging to bring about. Donor governments and the international community stand a better chance of "governing" migration in the context of climate change if it is said be a national rather than transnational concern.

45. Glen Coulthard, *Red Skin, White Masks: Rejecting the Colonial Politics of Recognition* (Minneapolis: University of Minnesota Press, 2014).

46. Ibid.

47. Fanon, *The Wretched.*

48. United Nations High Commissioner for Refugees, *Convention and Protocol Relating to the Status of Refugees* (Geneva: United Nations High Commissioner for Refugees, 1951), Art.1(2a), 14. This interpretation was upheld in 2015 by the New Zealand Court of Appeal, when it ruled that Ioane Teitiota's application for refugee status "on the basis of changes to his environment in Kiribati caused by sea-level-rise associated with climate change" could not be granted since "environment" was not among the criteria used to grant refugee status under the Geneva Convention. See Kelly Buchanan, "New Zealand: 'Climate Change Refugee' Case Overview," ed. The Law Library of Congress: Global Legal Research Center (Washington, DC: The Law Library of Congress, 2015).

49. Biermann and Boas, "Preparing for a Warmer World."

50. Ibid., 67.

51. Du Bois, *The Souls of Black Folk*, 11. On the hierarchy of the "color-line" and development, see, for example, Branwen Gruffydd Jones, "Race and the Ontology of International Relations," *Political Studies* 56, no. 4 (2008); Branwen Gruffydd Jones, "'Good Governance' and 'State Failure': The Pseudo-Science of Statesmen in Our Times," in *Race and Racism in International Affairs*, ed. Alexander Anievas, Nivi Manachanda and Robbie Shilliam (London: Routledge 2015).

52. Hobson, *Eurocentric Conception of World Politics.*

53. Carol Farbotko and Heather Lazrus, "The First Climate Refugees? Contesting Global Narratives of Climate Change in Tuvalu," *Global Environmental Change* 22, no. 2 (2012); Carol Farbotko, "'The Global Warming Clock Is Ticking So See These Places While You Can': Voyeuristic Tourism and Model Environmental Citizens on Tuvalu's Disappearing Islands," *Singapore Journal of Tropical Geography* 31 (2010); Carol Farbotko, "Wishful Sinking: Disappearing Islands, Climate Refugees and Cosmopolitan Experimentation," *Asia Pacific Viewpoint* 51, no. 1 (2010); McNamara and Gibson, "'We Do Not Want to Leave Our Land.'"

54. Carol Farbotko, "Tuvalu and Climate Change: Constructions of Environmental Displacement in Sydney Morning Herald," *Geografiska Annaler: Series B Human Geography* 87 (2005); Farbotko, "Wishful Sinking."

55. Epeli Hau'ofa, "Our Sea of Islands," *The Contemporary Pacific* 6, no. 1 (1994); Farbotko, "Tuvalu and Climate Change," 288.

56. Farbotko and Lazrus, "The First Climate Refugees"; R. McLeman, M. Moniruzzaman, and N. Akter, "Environmental Influences on Skilled Worker Migration from Bangladesh to Canada," *The Canadian Geographer* 62, no. 3 (2018); Robert Stojanov et al., "Local Perceptions of Climate Change Impacts and Migration Patterns in Malé, Maldives," *The Geographical Journal* (2016).

57. Farbotko and Lazrus, "The First Climate Refugees."

58. Allen Pred, *Even in Sweden: Racisms, Racialized Spaces, and the Popular Geographic Imagination* (Berkeley and London: University of California Press, 2000); Himani Bannerji, *The Dark Side of the Nation: Essays on Multiculturalism, Nationalism and Gender* (Toronto: Canadian Scholar's Press, 2000); Wendy Brown, *Regulating Aversion: Tolerance in the Age of Identity and Empire* (Princeton, NJ: Princeton University Press, 2006); Eva Mackey, *The House of Difference: Cultural Politics and National Identity in Canada* (Toronto: University of Toronto Press, 2002); Sunera Thobani, *Exalted Subjects: Studies in the Making of Race and Nation in Canada* (Toronto and Buffalo, NY: University of Toronto Press, 2007).

59. Wendy Brown, "Climate Change, Democracy and Crises of Humanism," in *Life Adrift: Climate Change, Migration, Critique*, ed. Andrew Baldwin and Giovanni Bettini (London: Rowman and Littlefield, 2017), 38.

60. Oli Brown, "The Numbers Game," *Forced Migration Review* 31 (October 2008); Norman Myers, "Environmental Refugees: An Emergent Security Issue" (World Economic Forum, Prague, 22 May 2005); Stern, *Stern Review*; Asian Development Bank, *Addressing Climate Change and Migration in Asia and the Pacific* (Manila: Asian Development Bank, 2012); Aid, *Human Tide*; Camillo Boano, Roger Zetter and Tim Morris, *Environmentally Displaced People: Understanding the Linkages between Environmental Change, Livelihoods and Forced Migration* (Refugee Studies Centre, Oxford Department of International Development, University of Oxford, 2008); Rigaud et al., *Groundswell*; Robert Nicholls, "Coastal Flooding and Wetland Loss in the Twenty-First Century: Changes under the SRES Climate and Socio-Economic Scenarios," *Global Environmental Change* 14, no. 1 (2004).

61. Biermann and Boas, "Preparing for a Warmer World."

62. Francois Gemenne, "Why the Numbers Don't Add Up: A Review of Estimates and Predictions of People Displaced by Environmental Changes," *Global Environmental Change* 21, no. S1 (2011); Biermann and Boas, "Preparing for a Warmer World"; Cord Jakobeit and Chris Methmann, "'Climate Refugees' as Dawning Catastrophe? A Critique of the Dominant Quest for Numbers," in *Climate Change, Human Security, and Violent Conflict: Challenges for Societal Stability*, ed. Jürgen Scheffran et al. (Berlin: Springer, 2012); White, *Climate Change and Migration*; Elizabeth Marino, "The Long History of Environmental Migration: Assessing Vulnerability Construction and Obstacles to Successful Relocation in Shishmaref, Alaska," *Global Environmental Change* 33 (2012).

63. Biermann and Boas, "Preparing for a Warmer World," 32.

64. François Gemenne, "Climate-Induced Population Displacements in a 4 Degree Celsius+ World," *Philosophical Transactions of the Royal Society London:*

*Mathematical, Physical and Engineering Sciences Series A* 369, no. 1934 (2011); Jane McAdam and Ben Saul, "Displacement with Dignity: International Law and Policy Responses to Climate Change Migration and Security in Bangladesh," *German Yearbook of International Law* 53 (2010).

65. Black, "Environmental Refugees."

66. McAdam and Saul, "Displacement with Dignity"; Richard Black, Dominic Kniveton and Kerstin Schmidt-Verkerk, "Migration and Climate Change: Towards an Integrated Assessment of Sensitivity," *Environment and Planning A* 43, no. 2 (2011), 432.

67. Brian Massumi, "Potential Politics and the Primacy of Preemption," *Theory and Event* 10, no. 2 (2007).

68. Gilles Deleuze, *Difference and Repetition* (New York: Columbia University Press, 1994); Rob Shields, *The Virtual* (London and New York: Routledge, 2003).

69. Amitav Ghosh, *The Great Derangement: Climate Change and the Unthinkable* (Chicago, IL: University of Chicago Press, 2016).

70. Warren, *Ontological Terror*, 6, my emphasis.

71. Ibid., 9.

72. Robert Kaplan, "Waterworld," *Atlantic Monthly*, 2008.

73. Massumi, "National Enterprise Emergency."

74. Jane Gordon and Lewis Gordon, *Of Divine Warning: Reading Disaster in a Modern Age* (Boulder, CO: Paradigm Publishers, 2009).

75. Michel Foucault, *Abnormal: Lectures at the Collège de France, 1974–1975* (New York: Picador, 2003) 56.

76. Ibid., 55–57.

77. Ibid., 317.

78. Gordon and Gordon, *Of Divine Warning*.

79. See also Gaia Giuliani, *Monsters, Catastrophes and the Anthropocene: A Postcolonial Critique* (London: Routledge, 2020).

# Chapter 3

# White World Order

Concerning the period after World War II when Britain was divested of its overseas territories, Bill Schwarz writes that "the presentiment of disorder that decolonization incubated was above all about race, driven by the fear that white ascendency was about to be turned upside-down, and that, in the final reckoning, the natives were to have their day."[1] Here Schwarz provides a glimpse into the affective power of racism. The "presentiment of disorder" he describes—a feeling, a brooding, a worry—was a racist anxiety, provoked when those colonised by the British asserted their political subjecthood in a direct challenge to British imperialism. Schwarz's observation is important because it reminds us that white racism is never reducible to an overt hatred directed towards its other but operates just as powerfully through affect and sense. The coloniser worries about the native's insubordination. The master lives in perpetual fear of revolt. But his observation is doubly significant for it also calls attention to the inherent instability of white racial rule. If decolonisation incubated a white racist anxiety, Schwarz reminds us that other conjunctures that threaten to overturn white rule can incubate a similar anxiety. It suggests, furthermore, that white ascendency's hold over its racial object is only ever a tenuous achievement and that maintaining this hold is internal to the logics and rationalities of white racial rule. It even suggests that a presentiment of disorder may be endemic to *all* expressions of white ascendency, that the fantasy of white racial rule is sustained by the persistent worry that one day it might all come apart.

The French and Haitian Revolutions certainly provoked such anxiety among the English landed gentry and aristocratic class. As Robbie Shilliam has shown, English elites were always wary that "the self-authored freedom of the enslaved would, in the absence of proper instruction, threaten to unleash a mass of indolence, licentiousness and anarchy."[2] This, they feared, threatened to undo the very patriarchal order that held the English "genus" together. John Hobson documents a similar racist anxiety in the geopolitical imaginary of the British geographer Halford Mackinder for whom the persistent threat of racial invasion along Europe's eastern border would galvanise the West's "defensive unification."[3] Or when Michel Foucault writes of National Socialism that "exposing the entire population to universal death was the only way it could truly constitute itself as a superior race," he offers the powerful insight that the fantasy of Aryanism was nurtured in the inverse fantasy of its complete annihilation.[4] What each of these examples shows is that when threatened, white racial rule redoubles the violence it directs towards its racial other. Extrapolating from these historical analogues, we might say that the violence of white ontology originates in the ever-present terror that the racial other—the colonised, subjugated, enslaved, the Muslim, the migrant or the refugee—might one day seek to rule over white people. Safeguarding against such ontological terror, thus, appears to have been a persistent feature of white European rule from at least the seventeenth century.[5]

I begin with these reflections on Schwarz's presentiment of disorder because a parallel can be drawn between it and the discourse on climate migration. Just as the period of decolonisation incubated white anxiety, so climate change, when conceived as a migratory crisis, can be understood to incubate a similar concern. The historical conditions that led to European decolonisation are, of course, vastly different from what we are experiencing today with climate change. Yet for many the anxiety is much the same. Here again are John Wennersten and Denise Robbins declaring the racial emergency of climate change: "Climate refugees, once thought to be a problem confined to segments of the developing world are on the verge of becoming a global problem. How we respond to the problem will dictate how well we can sustain ourselves in what we call civilized human society."[6] For these writers, climate change threatens to overturn an enduring racial hierarchy in which "civilized human society"—for them, the West—is understood to govern over the majority world. Theirs is a presentiment of *global* racial disorder, incubating in climate change, at the core of which is a desire to preserve what is imagined to be the primacy of the white Euro-western human ("how we sustain ourselves in what we call civilised human society"). It is precisely the desire to preserve this perceived global racial order sustained by the fantasy of its undoing that, I argue, is foundational to the logic of racial futurism so central to the adaptation of the political discussed in chapter 1.

Geoff Mann and Joel Wainwright argue that the political refers to "the ground upon which the relation between the dominant and dominated is worked out."[7] For these writers, to comprehend how capitalist political economy adapts to climate change means paying close attention to how the political itself adapts. This is a pivotal insight for any radical politics of climate change. Not only does it demand attention to the way in which structural inequality shapes the political landscape of climate change alongside fossil fuels and energy transition. It also demands that we consider how existing relations of political domination are maintained as a necessary condition for the adaptation of capital. My contention is that the discourse on climate change and migration is one such relation of domination through which the adaptations of the political can and should be traced. Following Giovanni Bettini, this means moving beyond the simplistic assumption that the climate change–migration relation is political because it is framed differently by different, often competing, interests.[8] Indeed, to make this assumption presupposes that the relationship between climate change and migration is an objective reality which is then simply subject to myriad different frames (threat, victim, etc.). But the political is far more fundamental than simply the fact that an object or a relation can be subject to competing framings. Instead, to truly understand the *political* nature of the relation requires that we scrutinise the very terms by which the climate changel–migration relation is called into being as a relation, asking not only how the relation is rendered meaningful but also who has the power to make it so. It means examining the conditions under which these terms are worked out, what their consequences are and for whom. Chapter 2 began the task of answering these questions by examining how the figure of the climate migrant/refugee might be understood in terms of anti-Blackness. In doing so, chapter 2 positions our discussion firmly within the political register of race and racism. The present chapter proceeds along similar lines, although here the aim is to grasp how the discourse functions as an anti-Black technology in a *planetary* context in which climate change threatens to undermine Euro-western humanism. The chapter examines the logics of racial futurism by interrogating how Euro-western humanism mobilises a presentiment of *international* climatic-migratory disorder as a technology of anti-Black power. Here, emphasis falls less on how the other of climate change is represented and more on what the international logics and rationalities used to represent the climate-migration relation can reveal about the whiteness of the discourse.

## RACE AND THE POSTCOLONIAL WHITE INTERNATIONAL

From its inception in the mid-1980s, with the publication of Essam El-Hinnawi's *Environmental Refugees* by the United Nations Environment Programme, the discourse on global environmental change and migration has been conceptualised primarily as an international problem and, therefore, one that ought to be governed by a mix of international institutions.[9] To better grasp this problematisation, I suggest we approach the discourse on climate change and migration as an example of a global racial project. Michael Omi and Howard Winant define a racial project as "simultaneously an interpretation, representation, or explanation of racial dynamics, and an effort to reorganize and redistribute resources along particular racial lines."[10] For Omi and Winant, racial projects are arrangements of power that "do the ideological work" of linking "structure and representation" to produce what they call racial formations. Racial formations are, for them, not accidents of history but conscious programmes that attempt to govern social relations racially. We can think of "climate change and migration" in much the same way. The assemblage of discourses, institutions and practices that problematise climate change as a pending migratory crisis can be understood as a specific *global* racial project. This is a racial project in which white racial rule adapts to the global climate change by mobilising both the structure of the international (e.g., international institutions, international law, regimes of knowledge production, racial capitalism) *and* a set of representational practices associated with international relations (IR), alongside those discussed in the previous chapter. Approaching the discourse as a specific kind of global racial project is useful for theorising racial futurism. Not only will doing so further clarify the racism at stake in this planetary topology. It also allows us to grasp how the white Euro-western human responds to the ontological terror of climate change. Here, the global racial project of climate change and migration shares much in common with the ontology of white rule. If climate change threatens white rule, global migratory disorder stands in for climate change as that which must be governed. When interpreted this way, we can see how the white Euro-western subject doubles down on the racial other of climate change as a matter of its own survival. My concern in this chapter is to trace how the logics of racial futurism take shape in relation to the concept of the international.

Measures devised by international institutions for governing so-called climate migrants/refugees are, therefore, of secondary interest to me in this chapter. More important is the work the category of "the international" accomplishes in sustaining the global racial project of climate change and migration. In fact, despite the discourse's growing popularity, international institutions have taken surprisingly few concrete measures to manage so-called climate change–induced migration.[11] International institutions, like the Nansen Initiative and its successor the Platform on Disaster Displacement, have certainly done much to frame the relation between climate change and migration as one of international law, in the language of human rights, for example. It is also true that numerous measures have been *proposed*, ranging from catastrophe bonds and parametric insurance to circular labour migration schemes, some of which are the focus of chapter 6.[12] Yet at the time of writing, the UNFCCC has taken no concrete measures aimed at governing the migratory effects of climate change aside from provisions in the Paris Agreement to strike a task force that would seek to better comprehend these effects in relation to the very specific category of human displacement. Indeed, on this point, the only consistent policy taken by the international climate change policy community in relation to migration is one in which parties to the convention have called on the research community for more research.[13] Driving this policy is a concern that knowledge about these effects is insufficient from the perspective of policymaking.[14] But here, once again we encounter the structuring effect of the onto-epistemic disjuncture referred to in chapter 1—governing migration in the context of climate change requires knowledge of a relation defined by its unknowability, which, as Sarah Nash has shown, results in a seemingly endless paradox: the search for a knowledge with no definitive object.[15]

To understand the ideological work accomplished by the international, the chapter approaches the international as a political resource, as opposed to a neutral set of institutions over and through which national governments and non-governmental organisations engage in international politics. Specifically, I approach the international as a category which furnishes the discourse with one of its principal meanings. The chapter draws, in particular, from the field of postcolonial IR theory, which challenges many of the taken-for-granted assumptions that shape both international society (e.g., the United Nations, intergovernmental institutions, nation- states and non-governmental organisations) but also the field of IR (e.g., academic knowledge), which takes international society as its object.[16] The argument relies especially on critical investigations that denaturalise the international by revealing its underlying Eurocentrism. One important feature of this critique is the claim that IR has a propensity to represent its object—the modern world order of nation- states—in a way that masks its origins in European imperialism.[17] IR is, thus, said to be constituted by forms of knowledge that must be forcibly removed from the dominant repertoire of IR in order for IR to cohere as a system of thought. Boundaries, in other words, must be erected around the field of IR for the field to make any meaningful sense to its adherents, boundaries that distinguish, for example, between permissible and impermissible topics for investigation within the disciplinary apparatus of IR. As we will soon see, the racist and colonial history of the international is one such impermissible topic.

In a direct challenge to the boundary-making practices of IR, Sanjay Seth argues that "any satisfactory account of the emergence of the modern international system cannot chart how an international society that developed in the West radiated outwards, but rather needs to explore the ways in which international society was shaped by the interaction between Europe and those it colonized."[18] For Seth, such a relational account of international society would seek to recover the ways in which colonialism, including its inflections of race, gender, class and sexuality, functions as a constitutive, even if disavowed, element of mainstream international theory and practice.[19] Taking up Seth's challenge, many have begun excavating how race and racism were central to the disciplinary mainstream of early IR theory and practice. Robert Vitalis has argued, for example, that IR emerged in the early twentieth century as a policy science that sought to consolidate Anglo-American world dominance in response to liberation struggles inspired by Pan-Africanism and Black internationalism, both of which, he argues, sought to overturn "the theory and practice of white supremacy." But as Vitalis puts it, many of the "intellectuals, institutions and arguments that constituted international relations were shaped by and often directly concerned with advancing strategies to preserve and extend that hegemony against those struggling to end their subjection."[20] Indeed, as Vitalis and many others have noted, one telling moment in this history is the fact that the now famous IR periodical *Foreign Affairs* was once originally the *Journal of Race Development*. It is the ideological cleansing of race and colonialism from the international that interests me here.

We can think of contemporary IR as an historically situated epistemological system that takes as its object the postcolonial international order. That is, IR is here conceived as a regime of power/knowledge, the purpose of which is to contain the excesses of the postcolonial order it seeks to govern. For clarification, my reference to postcolonial order follows a common line of argument in postcolonial theory, which is that the postcolonial designates the persistence of colonial relations even after formal systems of colonialism have come to an end.[21] By this account postcolonial order is merely the repetition rather than the passing of colonialism. At the heart of my argument in this chapter is the idea that the "presentiment of disorder," the racist anxiety that Schwarz found animating the decolonial period that began after World War II, continues to organise IR's postcolonial imaginary in the present. Or more precisely, a presentiment of disorder is foundational to the epistemological composition of IR. This is the idea that from the perspective of Europe, the postcolonial international order is always at risk of coming undone and that its potential undoing is also the potential undoing of a particular Eurocentric order of global governance.

There is no shortage of historical analogues one can use to draw links between the presentiment of racial disorder that prevailed during the twilight of the British Empire and those formative of nineteenth- and early twentieth-century IR. In their treatment of the global colour line, Lake and Reynolds tell the story, for example, of how the colonial authorities in Australia imposed immigration restrictions on Chinese migrant workers in order to protect white workers and mining investment from what they felt was unfair competition, restrictions that would later become central to the White Australia policy.[22] Similar immigration restrictions were also imposed in California and British Columbia, again in the interest of securing white labour markets from Chinese workers. Indeed, throughout the late nineteenth and early twentieth centuries "The West" was often explicitly referred to as the "White world." In fact, a growing, and often explicit, unease prevailed across the so-called White world, which held that, unless restrictive measures were taken to control immigration and reassert the imagined global dominance of the West, world white supremacy would be overturned by the so-called non-white people from Latin America, Africa and Asia. The surprise defeat of the Russians to the Japanese in the Russo-Japanese War in 1905, for example, became an emblem of what might happen to the "White world order" if measures were not taken to bolster European and American imperialism abroad and territorial boundaries at home. Similar sentiment is also evident in Lothrop Stoddard's popular 1920 book *The Rising Tide of Color: The Threat against White World- Supremacy*, in which the American author took pains to describe how the violent confrontation of European nations during World War I had greatly weakened the cause of world white supremacy. Although no imperialist, Stoddard would advocate immigration restrictions of non-white peoples to the "White world" as a eugenic imperative.[23]

Charles Pearson's widely read book *National Life and Character: A Forecast* told a similarly anxious story about the demise of world white supremacy. Pearson's text would also go on to play an outsized role in the political imagination of Theodore Roosevelt. But we could also add to this list Oswald Spengler's *The Decline of the West*, Madison Grant's *The Passing of the Great Race* and Maurice Muret's *The Twilight of the White Races*.[24] Vitalis documents how IR began to take shape as an academic discipline during this period in response to these narratives of global white decline. It was hoped that the new social science of IR would be used to forge a transnational Western alliance capable of ensuring the continued dominance of the "civilised world," that is, the West, in world affairs. By 1966, this expression of the international had come to be viewed by delegates at the Tricontinental Conference in Havana as the "White International."[25]

This is, of course, a very potted history. It, nevertheless, highlights how IR came into being during a period when racist geopolitical anxieties about the instabilities of White racial rule were explicit in the rhetoric of Western statesmen. It also further contextualises Schwarz's presentiment of disorder, which finds its historical antecedent in this period. Keeping this history in mind is important, then, when confronting how climate migration discourse functions as a global racial project within the epistemological framework of IR. Because even while considerable pains are taken to distance the discourse from the history of European colonial history, a point I will elaborate on below, we know from postcolonial studies that the main pillar of colonial thought—race—continues to organise postcolonial discourse even after formal colonialism. So, for example, we know that in postcolonial IR the colonial difference between the "developed" and the "developing" worlds signifies a racial distinction even if race is never immediately referenced. So, too, we know that the term "the Global South" refers to the Black and brown majority world, while "donor" or "donor community" signify whiteness. In this way we can conceptualise the epistemological system of postcolonial IR as a representational schema that reinforces the assumed dominance of whiteness, in part, through a commitment to colour-blind international theory.

## CLIMATE CHANGE, MIGRATION AND THE WHITE POSTCOLONIAL INTERNATIONAL

The remainder of the chapter explores how the white Euro-western human mobilises the postcolonial white international to secure itself from what it perceives as the pending global racial disorder incubating in the crisis of climate change. It examines, in particular, three epistemological motifs that obscure the raciological origins of the postcolonial white international: empiricism, historiography and planetary consciousness. These, I suggest, are just a few of the norms of the international that can help us grasp how the discourse on climate change and migration functions as a global anti-Black technology. Each is discussed in turn.

### Empiricism

Race and racism are notoriously overlooked categories in mainstream IR. For Branwen Gruffydd Jones, "part of the reason [for this] arises from the dominance of empiricism in mainstream IR."[26] In other words, mainstream IR produces knowledge about its object—international society—using the method of direct and immediate observation. This results in two problems for an analytic of race in IR. The first is that, because of its empiricist bent, IR confines its understanding of world order to only those phenomena it is capable of discerning. This includes racism, but only those instances of racism that can be discerned through empiricism, such as specific instances of racist discrimination perpetrated by individual nation- states, organisations or individuals, for example, those of a racist political leader. The trouble, however, is that by constructing racism solely as an objective phenomenon, as opposed to a structure or relation of power, mainstream IR remains unable to account for the structural racisms that underpin those acts of racism it *can* discern. As Stuart Hall makes clear, "Hateful as racism may be as a historical fact, it is nevertheless also a system of meaning."[27] But since empiricist methodology is unable to discern systems of meaning (such systems remain beyond the grasp of immediate observation), mainstream IR simply disregards any meaningful engagement with this important form of structural racism. Instead, mainstream IR constructs its object—international society—as though it were merely the anarchic interactions of discrete nation- states, as opposed to a spatial hierarchy, materialised by a representational system for the purpose of maintaining a Eurocentric world order. For Jones, empiricism is simply unable to grasp this unique attribute of international racism. Thus, the crux of her argument is that by reducing its conception of racism to empiricism, mainstream IR overlooks the ways in which both itself, as an epistemological system, and the international society it reflects reinforce global racial hierarchies between, for example, North/South, developed/developing, white/Black. Through its adherence to empiricism, IR obscures the role it plays in linking structure and representation to produce international society as a global racial formation, and in so doing, IR reinforces global structural racisms even as it seeks to reverse global inequality through undertakings such as the Sustainable Development Goals.

The second problem that empiricism generates for an analytic of race in IR stems from IR's emphasis on the individual state as its main unit of analysis. Using the concept of the failed state to illustrate her point, Jones shows how abstract metrics, like per-capita gross domestic product, the corruption index, the human development index and so forth, are used to assess the economic and political performances of states in international society.[28] Those that meet a specified threshold are then said to be "failed states."[29] The problem with this kind of state-level empiricism, however, is that it reduces the state to a set of indicators which has the effect of isolating the state from its concrete historical circumstances. This is especially problematic for Europe's former colonies, since the history of colonial state formation is assumed to bear no significant relationship with present-day state failure. By this reasoning, Haiti is often considered to be a failed state. But it gains this designation not because of Europe's refusal to recognise it after the successful slave revolution in the early nineteenth century, nor because of two decades of U.S. military occupation in the first half of the twentieth century, but because Haiti itself has failed to conform to a set of abstract, Eurocentric norms. The historical fact of racism is conveniently omitted from such state-centric empiricism.

Barnor Hesse refers to this representational habit as "epistemological racialisation," "the codified organisation of knowledge (e.g. deliberations, expertise, histories, representations, and explanations) based on the adjudication and valorisation of Europeanness and the debasement and appropriation of non-Europeanness, *but without explanatory reference to the impact of coloniality.*"[30] Charles Mills describes it as "the retrospective whiting out, the whitewashing, of the racial past in order to construct an alternative narrative that severs the present from any legacy of racial domination."[31] At its simplest, by decontextualising the states it describes, this kind of state-level empiricism contributes to the epistemological racialisation of IR. It whitens IR, if by whiteness we mean the power to occlude histories of racism from contemporary global consciousness.

An influential report published in 2018 by the World Bank titled *Groundswell: Preparing for Internal Climate Migration* exemplifies both problems.[32] Consider first the fact that *Groundswell* is a comprehensive empirical rendering of climate-migration dynamics, using what it calls "anticipatory diagnostics" to quantify "internal climate migrants" in three future scenarios (worst-case, moderate-case and best-case) projected out to 2050. By estimating the number of the so-called climate migrants that will materialise over the next three decades, its purpose is to assist decision makers in managing the disruptive effects of climate change on human population movements through planning, development and humanitarian intervention. Or to rephrase this slightly, its rationale is to contain the human excesses of climate change.

*Groundswell* is structured by a geographical focus on Latin America, sub-Saharan Africa and South Asia, which may seem to contradict Jones's point about the nation- state, but each region is carefully rendered empirically using graphs of projected climate migrants and time-sequence maps, while the report's most detailed scenarios are reserved for Ethiopia, Bangladesh and Mexico. To be sure, this is an international report produced by one of the world's most powerful international financial institutions about three specific regions and three specific states, each structured by differential conceptions of colonial difference. Each region and country is simply abstracted into a speculative empirical logic, which constructs each locale in relation to the future. Each place is said to be overfull with potential. Not overburdened by the past, the inhabitants of each location are said to be living on the verge of displacement. To point this out is not to suggest that the modelling techniques it uses are somehow erroneous or that the knowledge it generates is not robust. In fact, the report follows a rigorously delineated methodology. It is, however, to recognise that the knowledge it produces abstracts each country from the concrete historical reality of global inequality, about which writers like Jones, Vitalis and Seth tell us that it can and should be explained with reference to histories of colonial and racist dispossession. Indeed, to interpret these states without reference to these histories is precisely to engage in a form of epistemological racialisation that appropriates the lives of non-European people without reference to the impact of European imperialism.[33] The past invariably structures the everyday lives of those described in the text, and yet their pasts are said to be irrelevant to our understanding of migration futures. The authors of the text even tell us that "preexisting vulnerabilities shape the extent to which environmental change causes people to move and affects the type of movement." Yet no explanation is given as to the historical circumstances that might have brought these vulnerabilities into existence.

We can further recover the way in which race structures the meaning of *Groundswell*, if we consider, too, how it expresses a presentiment of racial disorder. Consider, for example, the report's title. According to the *Oxford English Dictionary* a groundswell is both "a deep swell or heavy rolling of the sea, the result of a distant storm or seismic disturbance" but also a form of "political agitation." *Groundswell* blurs these two meanings. It designates a coming political disturbance naturalised by the gathering forces of climate change in which the figure of the climate migrant—the racial other of climate change—is understood to threaten an existing political stability. Consider, too, how the text designates a presentiment of disorder not merely in reference to Ethiopia, Mexico and Bangladesh, but to the Black and brown majority world, a presentiment of racial disorder from the standpoint of the international. Importantly, however, the tone of the text is not one of anxious worry or fret; it is far more measured, more clinical, more empirical. A range of visual data aids appears throughout the text (e.g., graphs, tables, data) which gives the appearance that migration due to climate change is a manageable phenomenon; this is hardly an aesthetic rendering of racial ungovernability. But the text is equally clear that, if left unmanaged, these futures once materialised can have devastating consequences for migrants as well as for sending and receiving communities. The narrative tension that appears here between manageability and its excess recurs throughout the policy literature on climate migration, indicative of the pre-emptive logic at the heart of racial futurism. I address this logic in more detail in chapters 5 and 6. What concerns me here, however, is the way in which *Groundswell* cleanses this presentiment of disorder of its racial connotations. The futures described in *Groundswell* are entirely consistent with the explicitly racist futures described by Wennersten and Robbins. But in *Groundswell*, race is expunged, the discourse severed from any meaningful considerations of race and racism.

When we follow Jones, Seth and Vitalis, we are able to grasp how the discourse is purified of any obvious markers of racism when constructed as an international phenomenon. Racism may well be an attribute of individual nation- states, or indeed, specific international actors or individuals. But through its disavowal of European imperialism, as well as its adherence to empiricism, the epistemology of the international constructs international society as though it has transcended racism. Even while the figure of the climate migrant/refugee is racialised, and even while those said to be potential climate migrant/refugees are overwhelmingly coterminous with the Black and brown majority world, the international itself appears impertinent for understanding the discourse on climate change and migration as a global racial project. International society is understood to work implicitly in the background of climate migration discourse but is not itself understood as a European system of thought that invests the discourse with racial meaning. To some readers these issues of race and racism may seem unnecessarily complicated. But to approach the relation between climate change and migration as merely an empirical phenomenon, as many would insist we do, risks losing sight of not only how it is racialised, but even more importantly how the discourse itself functions as a frame of reference for understanding climate change. In turn, we lose sight of the way that it might also be understood as a means for shoring up the postcolonial white international amidst the crisis of humanism, or the crisis of whiteness that climate change can be understood to represent.

## Historiography

The second motif concerns the historiography of the field of environmental migration, which often includes climate change. By historiography, I refer to the histories told of the field, histories that seek to explain how the relation between climate change and migration comes into its own as an object of research. Scrutinising this historiography is important because doing so can help us grasp how the field is consolidated as a field. It allows us to trace the narrative contours of the field and, thus, to observe how these contours delimit permissible and impermissible lines of enquiry that define the field. In this way, we can understand historiography as central to the formation of epistemic communities, informal networks of actors bound together through shared assumptions about what counts as knowledge within a delineated field.[34] The argument I develop in this section is that "the international" figures prominently within the field's rather sparse historiography. Indeed, just as we saw with regard to empiricism, when the historiography of the climate changel–migration relation is narrated as an *international* relation, the "international" serves to mask the field's racist origins. That is, by constructing the climate changel–migration relation as international, the field's minimal historiography invites us to forget how contemporary knowledge about migration and environmental change is itself embedded within a racist genealogy traceable at least to the early nineteenth century. The space of the international enables this forgetting to occur.

How does this form of racial erasure work? To give an example, consider an argument made by Etienne Piguet in which he seeks to account for the emergence of "nature" or the "environment" as a legitimate category in contemporary migration studies.[35] Piguet explains that while "the environment" or "nature" were once popular categories in the foundational texts that would come to define migration research in the late nineteenth century, both slowly "disappeared" from migration research over the course of the twentieth century. For Piguet, these categories would then reappear in the 1980s such that both now enjoy renewed interest, even if this interest remains relatively marginal within the wider field of migration theory and practice. Piguet's interesting study merits our attention for at least two reasons. First, it is one of only a few that attempt to historicise the field of environmental migration within the broad trajectories of twentieth-century thought. What makes Piguet's intervention so interesting is that it scripts an intellectual history of the field of environmental migration which the field can then tell about itself. In this way, it supplies the epistemic community of "climate change and migration" with the means for constructing a shared intellectual identity from a shared history.[36] Robert McLeman and François Gemenne script a similar historiography in their introduction to the *Routledge Handbook on Environmental Migration and Displacement*. But the second reason Piguet's historiography is important is that it has acquired a kind of quasi-official status, given that it reappears, albeit in unattributed form, in the opening pages of *The Atlas of Environmental Migration*, published by the International Organization for Migration in 2017.[37] Thus, not only does it contribute to the formation of an epistemic identity, but it also shapes the wider international political and policy landscape within which the climate-migration relation emerges as an object of political reflection.

To understand how racial erasure works in Piguet's historiography, let us consider his argument in more detail. He begins first by establishing the fact that the "environment" played a formative role in the nascent field of migration studies in the nineteenth century. Tracing this early history to Moritz Wagner's theory of plant and animal migration, Friedrich Ratzel's *Anthropogeographie*, Ernst Georg Ravenstein's *The Laws of Migration*, environmental determinists such as Ellen Churchill Semple and Ellsworth Huntington, and the anarchist geographer Peter Kropotkin, Piguet argues that "early attempts at theorizing migration agreed on a central role for the natural environment."[38] He then gives four broad explanations to account for the disappearance of the environment in migration studies that would occur over the course of the twentieth century. First, as technology became more sophisticated, Western societies would become less constrained by "nature." Technological sophistication, such as automobility and rail, ensured that human mobility, including migration, was able to transcend the constraints imposed by nature. Second, he explains how environmental determinism came under heavy criticism both for "the naïvete of the monocausal explanations" it offered and for the way it had come to "legitimize a shamelessly racist conception of the world and a natural hierarchy of development to justify colonial oppression."[39] Third, he notes the increasing primacy migration scholars placed on the economic underpinnings of migration, and, fourth, that with the advent of forced migration studies, migration was increasingly said to result from political factors (e.g., state practices). At this point, it is important to note that none of these explanations are especially objectionable. We might quibble over precise details, but, in the main, Piguet provides a reliable explanation for the environment's disappearance from migration research. Piguet then attempts to explain why "the environment" *reappeared* as a topic within migration studies. The reasons he gives are increasing environmental consciousness in the West; rising incidents of environmental hazards; "growing anxiety about climate change"; increasing interest in climate change among social scientists; a new security paradigm after 9/11 that would link climate change to national security through the concept of "climate security"; a "disinhibited reading" of the human dimensions of environmental change research; and, finally, greater sensitivity within migration studies to the multicausal underpinnings of migration. Here, again, nothing on this list is especially objectionable. All of these historical events are certainly worthy of our consideration.

But if our task is to critically evaluate how the field's historiography draws boundaries around the field, then we also need to consider what world- historical phenomena are *missing* from Piguet's historiography. Although listing such omissions would be an endless task, for my purposes, one noticeable absence is any appreciation for the ways in which the political economic transformations brought about by decolonisation could explain the appearance-disappearance-reappearance narrative he tells. Piguet does, of course, acknowledge, albeit briefly, that environmental determinism was centrally important to sustaining racial hierarchies and colonialism. But it is hardly a stretch to suggest that he downplays the wider context of European imperial decline that shaped the political economic transformations of twentieth-century capitalism. Again, Piguet is quite right in saying that environmental determinism fell into disrepute because it was a naïve pseudoscience that underwrote capitalism, colonialism and National Socialism. So, my criticism here is not a matter of kind but of emphasis. Nor is it my intention to disparage Piguet nor his desire to recuperate the biophysical and geophysical materialities that shape the complex histories with which he is grappling. I would even hasten to add that his call to denaturalise nature and the environment in environmental migration research holds considerable merit. The problem, however, and this is actually of crucial importance, is that by neglecting to account for the broad sweep of decolonial history in his historiography of environment and migration research, he invites his readers to forget that the history of twentieth-century racism and Europe's postimperial context can be of vital importance in explaining why nature and the environment *re*surfaced.

By the mid-twentieth century the very intellectual scaffolding that supported the fantasy of European racial superiority over the preceding century and a half—including environment determinism, eugenics and racial science—had crumbled under the weight of its inherent violence, hastened in no small part by the horrific events of the Nazi Holocaust. One important effect of this collapse is that it gave rise across the West to what Sunera Thobani has called a "crisis of whiteness."[40] As European nation- states struggled to rebuild after the war, they found it increasingly difficult to govern their colonial territories. In 1947, Britain would finally withdraw from India, an event that would galvanise an already forceful anti-colonial movement that set in motion the decolonial struggles across what is today the Global South. But this withering of European colonialism would also expose a commensurate cultural frailty in Europe itself. Indeed, it was precisely during this period of white crisis, so thoroughly documented by Schwarz, that a presentiment of racial disorder incubated in Britain. This was the widespread anxiety that the British Empire was well and truly finished. Paul Gilroy calls this cultural logic of white loss "postcolonial melancholia."[41] Needless to say, far more could be said about this tumultuous period of European decline. Though, for our purposes, it is enough simply to name its omission from Piguet's historiography because it is precisely within this milieu of white loss that the environment's *reappearance* in migration research first began to stir. Piguet attributes the actual reappearance of the environment in migration studies to a series of international publications from the late 1980s. Patricia Saunders is very clear, however, that its origins rest in the Malthusian revival of the 1970s, notably the Worldwatch Institute and its founder Lester Brown,[42] a time when bloody decolonial battles were being waged in all corners of the Global South.

It would be simplistic to argue that 1970s global environmentalism amounted to the reassertion of whiteness in response to a globalised crisis of whiteness.[43] It would not be unreasonable, however, to suggest that this crisis can help to explain the reappearance of the environment in migration research. Yet this is precisely what Piguet does. In downplaying the colonial and imperial histories implicated in the environmental revival of the 1970s, his historiography is cleansed of any meaningful *continuity* with the racist origins of early migration research. Indeed, he overlooks one of the most significant political developments in twentieth-century history: the assertion of political subjecthood by the Black and brown majority world, evident not only in decoloniality and Third Worldism but also in the U.S. civil rights movement and in worldwide movements for Indigenous self-determination.

But equally important, and this brings us back to the question of the international, according to Piguet's historiography, when the environment *does* resurface in the postcolonial context of migration studies, it resurfaces *in the domain of the international*. That is, the global environmental crisis resurfaces, at least in part, as a pending *international migration* crisis. Specifically, he identifies how the environment-migration relation reappears in the postcolonial international context in a set of international policy texts, citing, in particular, Jodi Jacobson's Worldwatch Institute *Environmental Refugees: A Yardstick of Habitability*, published in 1988, and the 1990 report of the IPCC and Essam El-Hinnawi's *Environmental Refugees*, published by United Nations Environmental Programme in 1985. In other words, when the environment resurfaces as a means for explaining migration, it takes shape as a crisis of global governance specific to the Global South.

But what is additionally important here is that, according to Piguet, it is only *after* the publication of these international policy texts that the environment gets taken up as an object of academic investigation in contemporary migration studies, not before, and even then, only minimally. McLeman and Gemenne argue precisely the same sequence of events in their historiography in the *Routledge Handbook on Environmental Migration and Displacement*. In effect, what we have then is a situation in which international environmental actors are seen to be reconstructing the environmental crisis as a migration crisis, and *not* a case of migration scholars suddenly stumbling upon the environment as a forgotten category that needs to be resurrected. This may seem like a pedantic difference to some readers. Just consider though that the international is not epistemologically neutral but rooted in a European racial imaginary. As we saw earlier, it does important ideological work. It decouples the construction of modern international order from its foundations in race thinking, while the discipline of IR was itself developed as a policy science that would mitigate early twentieth-century anxieties about white decline through the management of colonial difference. That the environmental crisis comes to be constructed in the 1970s and 1980s as an *international* migration crisis in many ways repeats the founding logics of IR, where governing colonial difference is taken as a prerogative of "white world order." Only now it is not simply colonial difference that is subject to the rule of IR but the very migration of colonial difference.

All of this should give us pause to reconsider Piguet's historiography. In his historiographic representation of environmental migration research, race and white supremacy are assigned very minor roles and only then relegated to the very distant past where they can be safely viewed as unfortunate markers of a field that has formally, and thankfully, transcended its racial history. The result, I argue, is that through historiographies like Piguet's (but also in McLeman and Gemenne's) racism and white supremacy have been modernised out of the field, such that we end up with a colour-blind historiography of a field that, I would suggest, was only ever about race. Accordingly, the field comes to be delimited in such a way that race and racism fall outside the scope of its epistemic remit. The task remains, then, to read race back into the field, to understand the way the field operates as a global racial project.

## Planetary Consciousness

The final motif of the international is that of planetary consciousness. This is a distinctive geographical perspective, or viewpoint, that holds within its gaze the totality of the Earth and its inhabitants. Just as we saw with empiricism and historiography, deconstructing how planetary consciousness structures the discourse on climate change and migration can further clarify how the discourse functions as a system of racial meaning. As we will soon see, when constructed as an international crisis, knowledge about the climate-migration relation is very often expressed from the perspective of this totalising world view. Planetary consciousness hovers over the space of the international from a distance, imagining itself uniquely positioned both to narrate and to govern global relations. Within the discourse on climate change and migration, it problematises the migration effects of climate change as a pending global crisis, while holding that the international is key to its resolution.

Not only, then, does planetary consciousness allow the totality of "climate change and migration" to be imagined as a global problem, but it is also the subject position that claims responsibility for those imagined to be at risk of climate change. In today's hawkish political environment in which migrants are routinely vilified, this seems entirely justifiable; taking responsibility for the world's most vulnerable is hardly a contestable position, especially if undertaken in the name of climate justice. After all, as Patricia Noxolo, Parvati Ragurham and Clare Madge make clear, "Responsibility is increasingly summoned as a route to living ethically in a postcolonial world."[44] All the more so as the climate crisis becomes ever more palpable. But even as planetary consciousness may seem self-evident, it is also a powerful political ruse. Taking responsibility for the "victims" of climate change may seem to be an indisputable ethical imperative. But ethics are never so clear cut. "Responsible action," as Noxolo, Ragurham and Madge go on to elaborate, "is never free of its locational imperatives and its identifications so that the responsible agent is always tainted: there is no pure space within and from which responsibility can be enacted."[45] In other words, planetary consciousness is not some innocent imaginary. It, too, is inherently political.

What kind of political imaginary is planetary consciousness? Where does it originate? Detached from any particular location, nation- state or interest, planetary consciousness imagines itself capable of observing the world in its entirety from a singular perspective from above. Emblematic of this perspective is the iconic NASA photograph of the Earth from space,

AS17-148-22727, taken by the Apollo 11 space flight in 1972. AS17-148-22727 became instantly famous when it first appeared on Western television screens and print media. Looking back, it is easy to grasp the influence that AS17-148-22727 would have on various global imaginaries. Part of its appeal lay in the way it articulated what Tariq Jazeel has called "a new humanism." This was an image around which its mostly Western audience could unite to steward an ecologically vulnerable planet.[46] But not only would it come to shape what we know today as the global environmental movement, so it would also serve the ideological needs of what would later become globalisation. Nowadays, it is endlessly reproduced in contemporary media culture. We see it, for example, every time we use the popular Internet software Google Earth.

What makes this image so significant, however, is its extraordinary capacity to dissimulate its own conditions of production. It may well evoke the impression of an innocent Earth—the fragile "blue marble" suspended in the cosmos—outside the realms of politics or history. But it reflects something quite different. The locational imperative of AS17-148-22727 is firmly anchored to American technoscience and capitalism. What at first appeared transcendent, disembodied and decoupled from any immediate Earth-bound context turned out to reflect nothing more than raw imperial ambition, a totalising desire for unlimited control of Earth in the guise of responsibility, stewardship and care. In this way, the image can be reread as an assertion of U.S. imperialism. It may well have cultivated a sense of shared humanity around an optimistic faith in a future of American benevolence. What this masks, however, is that the image gained popularity at the moment when American geopolitical hegemony was anything but a certainty. America was losing the war in Vietnam where it had been embroiled for almost two decades. The Cold War was still in full force. While the twin effects of the 1973 oil shock and rapidly rising inflation would destabilise faith in American economic hegemony. In such a moment of crisis, it is hardly surprisingly, then, that AS17-148-22727 would be used to restore faith in American exceptionalism. Here was an image that offered to renew American innocence by coalescing the forces of capitalism, humanism, environmentalism and techno-optimism in the service of a fragile Earth. But what would appear purposeful and optimistic was ultimately a means of resolving the interlocking geopolitical crises of capitalism and whiteness. Donna Haraway would call attention to the situatedness of this perspective by famously renaming planetary consciousness the "god-trick of seeing everything from nowhere." For her, such an unmarked perspective amounted to "the power to see and not be seen, to represent while escaping representation," ultimately signifying "the unmarked positions of Man and White."[47] It is, perhaps, no coincidence, then, that the Worldwatch Institute founded by Lester Brown in 1974 was among the first to make reference to the figure of the environmental refugee.[48] Environmental worldwatching, it would seem, expressed a planetary consciousness that was tethered to its racial other from the outset.

Mary Louise Pratt describes planetary consciousness as a form of subjectivity foundational to modern European thought.[49] She traces its modern incarnation to the eighteenth-century natural historian Carl Linnaeus, who deployed it as a mechanism for constructing what she calls "global-scale meaning through the descriptive apparatus of natural history." Linnaeus is, of course, famous for his work *Systema Naturae* in which he developed a taxonomy for classifying plant life, laying the foundations for the study of natural history, including biology, geology and nineteenth-century racial science. Immanuel Kant's racial taxonomy would have been inconceivable without natural historians like Linnaeus and Buffon.[50] For Pratt, though, the overall significance of planetary consciousness found in the works of Linnaeus, Kant and others is that it sought to bring order to the natural world. It was a system of classification that wrestled the chaos and heterogeneity of the natural world into a singular system of thought in which all species of flora and fauna, including humans, were assigned a distinctive place in relation to all other species. Although European planetary consciousness is traceable to ancient Greece and Rome, its early modern expression took shape precisely through this arrangement of order. To be European in this natural historical sense was to watch over, preserve and gain mastery of the presumed order of nature on a planetary scale. Thus, planetary consciousness would henceforth not only "bear the burden of European thought and history," as Jazeel puts it.[51] So it would designate a perspective that would "measure, recognize and arbitrate on difference through the very categorizations it has conjured into existence."[52] It is no coincidence that Kant's racial taxonomy would inform the very same planetary self-conception of the European.[53]

That European planetary consciousness shapes the production of international knowledge about the climate migration is hardly surprising. The red arrows (discussed in chapter 2) bearing down on the West from the Global South, an allegory for racial invasion, are illustrative. The racial nomos implied by planetary consciousness is precisely what gives the arrows their meaning. They signify the very undoing of an order of European racial nomos, not unlike that fantasised by Stoddard or Wennersten and Robbins. But the raciality of planetary consciousness is not confined just to the realm of alarmist aesthetics. It is also endlessly reproduced in more formal strands of international knowledge about climate change and migration.

*The Atlas of Environmental Migration* published by the IOM reproduces a range of similar images from the very same planetary perspective.[54] In fact, the entire *Atlas* is narrated from this perspective. Dozens of images map the relation's multiple, intersecting dimensions. One such map, "The World in

Motion," maps global migration pathways where migration is said to be the "defining feature of world order."[55] *The Atlas* also attempts to map processes of return migration, immobility, relocation, forced and voluntary migration. But if *The Atlas* gives an overwhelming, even if simplified, account of migration, the presentiment of disorder it produces derives from the way it maps the planetary environmental crisis onto migration. Geophysical disasters, disaster risk, ecosystem degradation, droughts, wildfires, floods, climate risk, industrial accidents and land grabbing are all projected onto the world map in *The Atlas*. Together they invite the readers to imagine not just the world in its totality but to imagine the complex totality of environmental change and migration within the space of a singular text. In this sense, we can read *The Atlas* as a vision of total mastery—another episode in a much longer genealogy of European planetary consciousness which seeks to wrestle the disorder of an unruly, heterogeneous planet into some semblance of order from a distance. Indeed, the text's narrative arc represents environmental change and migration as a vastly complicated nexus of interconnecting social-environmental relations that calls for a logic of planetary governance, a crisis that *needs* to be intervened upon.

Yet what is so remarkable about *The Atlas* is that, even as it attempts to render governable the phenomenon it seeks to name, it struggles to discern that very phenomenon. Entirely consistent with the onto-epistemic disjuncture of climate-migration discourse—its undecidability, incalculability, ambiguity and so forth—*The Atlas* circles around the climate-migration relation without ever giving it objective form. With every layer of detail it adds about the environmental crisis, with every map, chart and graph, the very object of environmental migration becomes evermore elusive. It is almost as though a coherent grasp of the phenomenon "environmental migration" recedes further into the distance, the more the authors struggle to bring it into existence. In the early pages of *The Atlas*, the authors are very clear that "environmental migration" is a political construct, by which they mean that the concept "does not always match empirical reality," a concept "marked by many conflicting viewpoints and perspectives, reflecting different policy agendas."[56] A surplus of meaning to be sure. The text also endlessly reasserts the disclaimer that climate change should never be thought of as the cause of migration, displacement and resettlement, which, in turn, raises the important question as to what precise empirical reality it *does* seek to name. Even here we find the authors themselves frustrated by their own attempt to map the relation. It is worth quoting *The Atlas* at length:

> Independent of empirical reality these terms [i.e., environmental migrant] are also political constructs that are useful for highlighting the growing importance of environmental degradation as a factor of migration. It is not so much a matter

> of creating a particular category of migration as of drawing attention, as this Atlas does, to a neglected factor, whose importance will increase in the future.[57]

What kind of international knowledge is it whose sole purpose is merely to create the impression of a problem? How useful is this knowledge and for whom? But even as the authors attempt to resolve this conundrum by defining "environmental migration" using the IOM's expansive definition,[58] they are also very clear that the relation is beset by a range of unresolved methodological uncertainties, which, they suggest, can only be resolved when a firm decision is taken (by someone, anyone) since the relation is inherently political. Yet they go on to lament, "No one is really taking the risk of deciding either way, and everyone is satisfied with the ambiguity arising from a definition that is intentionally broad and flexible."[59] It is hard to reconcile the relation's undecidability that is so clearly evident in this section with the last section of *The Atlas*, which maps out a set of policy responses that its authors hope might be mobilised to address the multifaceted nature of the problem it struggles to define. Yet even here the sheer breadth of the problems brought under the sign of "environmental migration" undermine the very possibility that "it" might even be governable. The result is a speculative reading of a speculative phenomenon too excessive to define, let alone govern, yet too freighted with meaning not to.

How should we make sense of this authoritative text from the authoritative international body on migration? What starts out as an attempt to represent the full breadth of environmental migration from the heights of planetary consciousness ends up revealing far more about the locational imperative from which the text is written than it does "environmental migration." Even as it seeks to enumerate the latter, *The Atlas* expresses European planetary humanism in crisis. Hardly an optimistic account of the world, it reflects a calamitous planetary environment, not just fraying at the seams but a full-fledged coming apart of global nature. Whatever claim European planetary humanism imagined it might have had over the global environment in the heyday of global environmental governance seems now a distant memory. No traces of the heady optimism that accompanied the signing of the Rio conventions are to be found in *The Atlas*, only an appeal to manage the pending crisis of "environmental migration." No longer, it would seem, is planetary consciousness the source of a smooth techno-optimism or the rational ordering of the world unfolding as universal history. We are a long way from the promise vested in AS17–148–22727. In the intervening half century, the one-world ideology expressed in AS17–148–22727 is now giving way to a world under immense duress, a "savage ecology," reflected in *The Atlas* as a converging set of catastrophes that are simultaneously geopolitical and *geo*-political.[60] Here, the incalculability, undecidability and monstrous ambiguity of the figure of climate migrant/refugee is mirrored back to us in *The Atlas* as a crisis of planetary consciousness; *The Atlas*'s inability to represent the other of climate change mirrors its opposite: the withering perspective from which it is told. By attempting to tell a story about its other—the other of climate change—*The Atlas* inadvertently winds up telling a story about the undoing of a Eurocentric conception of the world, the undoing of a postcolonial white international once dominated by the West.

## CONCLUSION

If the political is about the relation between dominant and dominated, then the postcolonial white international is an historical manifestation of this relation spatialised globally as coloniser and colonised. This is the relation we need to grasp if we are to comprehend how migration structures the adaptation of the political to climate change. The postcolonial white international is a form of world order in which Europe's former colonies would acquire full-fledged statehood in the international system. But however much the violent anti-colonial struggles that marked this epochal transfer of power were vested with emancipatory promise, the resulting postcolonial white international would thereafter bear the legacy of colonial relations centred in Europe and the West. "Development" would come to organise how these newly "developing countries" figured within the global division of exploitation refracted, first, through the spatial logic of the Cold War and later, following the collapse of the Berlin Wall, through economic globalisation in which the United States became world hegemon. Accounting for this vital history remains an urgent task for coming to terms with migration in the context of climate change.

I have attempted to show in this chapter how the discourse of climate change and migration can be understood as a global racial project whose primary purpose is to recentre white Eurocentrism in the face of its undoing. This is a racial project brought into being and sustained by the international. I would argue that any attempt to grapple with the adaptation of the political must therefore grapple with the legacies of racism as they materialise the interface of migration, climate change and the international.

## NOTES

1. Schwarz, *Memories of Empire*, 29.
2. Robbie Shilliam, *Race and the Undeserving Poor: From Abolition to Brexit* (London: Agenda Publishing, 2017), 19.
3. Hobson, *Eurocentric Conception of World Politics*, 127.

4. Michel Foucault, *"Society Must Be Defended": Lectures at the Collège de France* (New York: Picador, 2003), 260.

5. Warren, *Ontological Terror.*

6. Wennersten and Robbins, *Rising Tides*, 32.

7. Mann and Wainwright, *Climate Leviathan.*

8. Bettini, "Climate Barbarians at the Gate?"

9. El-Hinnawi, *Environmental Refugees.*

10. Michael Omi and Howard Winant, *Racial Formation in the United States: From the 1960s to the 1990s*, third edition (New York: Routledge, 2015), 125.

11. It is worth noting that while several national governments have shown an interest in governing climate-related displacement, such as resilience funding provided to the Isle de Charles band by the U.S. Department of Housing and Urban Development and the migration with dignity policy taken by the Government of Kiribati, these policies are designed to assist the relocation of people in advance of climate change. They are in this sense preventative measures more than they are policies that govern climate migrants/refugees. For details on the case of Isle de Jean Charles, see Julie Maldonado and Kristina Peterson, "A Community-Based Model for Resettlement," in *The Handbook on Environmental Migration and Displacement*, ed. Robert McLeman and François Gemenne (Abingdon: Routledge, 2018), 289–99. On "migration with dignity," see Karen McNamara, "Cross-Border Migration with Dignity in Kiribati," *Forced Migration Review* 49 (May 2015).

12. Asian Development Bank, *Addressing Climate Change and Migration in Asia.*

13. Nash, *Negotiating Migration.*

14. This point is echoed by Robert McLeman and François Gemenne in their introductory chapter of the *Routledge Handbook of Environmental Migration and Displacement*. See Robert McLeman and François Gemenne, "Environmental Migration Research: Evolution and Current State of the Science," in *Routledge Handbook of Environmental Migration and Displacement*, ed. Robert McLeman and François Gemenne (Abingdon: Routledge, 2018), 3–16.

15. Nash, *Negotiating Migration.*

16. Seth, *Postcolonial Theory.*

17. Geeta Chowdhry and Sheila Nair, "Power in a Postcolonial World: Race, Gender, and Class in International Relations," in *Power, Postcolonial and International Relations: Reading Race, Gender and Class*, ed. Geeta Chowdhry and Sheila Nair (Abingdon: Routledge 2004).

18. Seth, *Postcolonial Theory*, 20.

19. Chowdhry and Nair, "Power in a Postcolonial World."

20. Vitalis, *White World Order*, 2

21. Edward Said, *Culture and Imperialism* (New York: Vintage Books, 1994); Derek Gregory, *The Colonial Present* (Oxford: Blackwell, 2004).

22. Lake and Reynold, *Drawing the Global Colour Line.*

23. Stoddard, *The Rising Tide.*

24. For an excellent account of these and other similar texts, see Musab Younis, "'United by Blood': Race and Transnationalism during the Belle Époque," *Nations and Nationalism* 23, no. 3 (2017). See Maurice Muret, *The Twilight of the White*

*Races* (New York: Charles Scribner's Sons, 1926); Oswald Spengler, *The Decline of the West* (London: George Allen and Unwin, 1932); Charles Pearson, *National Life and Character: A Forecast* (London: Macmillan and Co., 1893); Madison Grant, *The Passing of the Great Race* (New York: Charles Scribner's Sons, 1916).

25. Richard Drayton, "The Road to the Tricontinental: Anti-Imperialism and the Politics of the Twentieth Century" (*Legacies of the Tricontinental: Imperialism, Resistance, Law*, Centre for Social Studies, University of Coimbra, 22–24 September 2016).

26. Branwen Gruffydd Jones, "Race in the Ontology of International Order," *Political Studies* 56, no. 4 (2008): 909.

27. Stuart Hall, *The Fateful Triangle: Race, Ethnicity, Nation* (Cambridge, MA, and London: Harvard University Press, 2017), 33.

28. Jones, "'Good Governance' and 'State Failure.'"

29. It is noteworthy that while the concept of the "failed state" has been supplanted by that of the "fragile state," the latter is still made evident using empiricism.

30. Barnor Hesse, "Racialized Modernity: An Analytics of White Mythologies," *Ethnic and Racial Studies* 30, no. 4 (2007), 656.

31. Charles Mills, "Global White Ignorance," in *Routledge Handbook of Ignorance Studies*, ed. Matthias Gross and Linsey McGoey (London: Routledge, 2015).

32. Rigaud et al., *Groundswell.*

33. Readers should note that unlike Bangladesh and Mexico, Ethiopia was never colonised by a European power. Ethiopia, therefore, occupies a less certain relation to European imperialism than the other two.

34. Noel Castree, *Making Sense of Nature: Representation, Politics and Democracy* (Abingdon: Routledge, 2014); Peter M. Haas, "Introduction: Epistemic Communities and International Policy Coordination," *International Organization* 46, no. 1 (1992).

35. Etienne Piguet, "From 'Primitive Migration' to 'Climate Refugees': The Curious Fate of the Natural Environment in Migration Studies," *Annals of the Association of American Geographers* 103, no. 1 (2014).

36. Castree, *Making Sense of Nature.*

37. Dina Ionesco, Daria Mokhnacheva and François Gemenne, *The Atlas of Environmental Migration* (London and New York: International Organization for Migration/ Routledge, 2017).

38. Piguet, "'Primitive Migration' to 'Climate Refugees,'" 150.

39. Ibid., 151.

40. Thobani, *Exalted Subjects.*

41. Gilroy, *Postcolonial Melancholia.*

42. On the Malthusian inflections in climate migration discourse, see Bettini, "Climate Barbarians at the Gate?"; Giovanni Bettini, "(In)Convenient Convergences: 'Climate Refugees,' Apocalyptic Discourses and the Depoliticization of Climate-Induced Migration," in *Deconstructing the Greenhouse: Interpretive Approaches to Global Climate Governance*, ed. Chris Methmann, Delf Rothe and Benjamin Stephan (Abingdon: Routledge, 2013); Betsy Hartmann, "Rethinking Climate Refugees and Climate Conflict: Rhetoric, Reality and the Politics of Policy Discourse," *Journal of International Development* 22 (2010).

43. Discussion around this theme is taking shape in relation to the Anthropocene. See, for example, Timothy W. Luke, "Tracing Race, Ethnicity, and Civilization in the Anthropocene," *Environment and Planning D: Society and Space* 38, no. 1 (2020); Zoe Todd, "Indigenizing the Anthropocene," in *Art in the Anthropocene: Encounters among Aesthetics, Politics, Environments and Epistemologies*, ed. Heather Davis and Etienne Turpin (London: Open Humanities Press, 2015); Nicholas Mirzoeff, "It's Not the Anthropocene, It's the White Supremacy Scene; or the Geological Color Line," in *After Extinction*, ed. Richard Grusin (Minneapolis: University of Minnesota Press, 2018); Yusoff, *A Billion Black Anthropocenes or None*.

44. Patricia Noxolo, Parvati Ragurham and Clare Madge, "Unsettling Responsibility: Postcolonial Interventions," *Transactions of the Institute of British Geographers* 37, no. 3 (2012): 418.

45. Ibid., 42.

46. Tariq Jazeel, "Spatializing Difference beyond Cosmopolitanism: Rethinking Planetary Futures," *Theory, Culture & Society* 28, no. 5 (2011).

47. Donna Haraway, "Situated Knowledges: The Science Question in Feminism and the Privilege of Partial Perspective," *Feminist Studies* 14, no. 3 (1988), 581.

48. Saunders, "Environmental Refugees."

49. Mary Louise Pratt, *Imperial Eyes: Travel Writing and Transculturation* (London: Routledge, 1992).

50. Jon Mikkelsen, ed., *Kant and the Concept of Race: Late Eighteenth-Century Writings* (Albany: State University of New York Press, 2013).

51. Jazeel, "Spatializing Difference," 85.

52. Ibid., 85.

53. Mikkelsen, *Kant*.

54. Ionesco, Mokhnacheva and Gemenne, *The Atlas of Environmental Migration*.

55. Ibid., 8.

56. Ibid., 6.

57. Ibid., 3.

58. Ibid., 3. According to the International Organization for Migration, "Environmental migrants are persons or groups of persons who, predominantly for reasons of sudden or progressive change in the environment that adversely affects their lives or living conditions, are obliged to leave their habitual homes, or choose to do so, either temporarily or permanently, and who move either within their country or abroad." This is an incredibly expansive definition that can be interpreted to cover almost every conceivable form of migration. Its meaning hinges on what is meant by "predominantly," "lives," "living conditions" and "environment." When I read this definition I struggle to see how it can be used to discriminate between those made homeless by the 2008 financial crisis and those made homeless by extreme weather events in coastal Bangladesh.

59. Ibid., 3.

60. Grove, *Savage Ecology*.

## Chapter 4

# From Determinism to Complexity

In his best-selling account about growing up Black in America, *Between the World and Me*, Ta-Nehisi Coates writes, "My experience in this world has been that people who believe themselves to be white are obsessed with the politics of personal exoneration."[1] For Coates, those who identify as white often go to great lengths to distance themselves from any personal implication in historical and contemporary forms of racism. White exoneration is also the premise of Reni Eddo-Lodge's *Why I Am No Longer Talking to White People about Race*. "I can't talk to white people about race anymore," she writes, "because of the consequent denials, awkward cartwheels and mental acrobatics when [racism] is brought to their attention."[2] Although important differences separate these authors, both pinpoint how the attempt to exonerate oneself from racism has become an important hallmark of contemporary whiteness. Alana Lentin describes this peculiar condition as "not racism."[3] For Lentin, "not racism" is a distinctive form of contemporary racist violence in which the "deepening and expansion of systemic, state and popular racism against migrants and asylum seekers, the undocumented, Indigenous people, Muslims and Black people" is accompanied "by an ever more vigorous denial that these phenomena are racist."[4] White exoneration is also central to David Theo Goldberg's conception of the postracial—the paradoxical condition in which racism becomes fully expressed in its denial.[5] For all four writers, the "move to innocence" associated with declarations of white exoneration is a key pivot in the consolidation of contemporary racism, amounting not to the transcendence of racism but to its intensification.[6]

I begin this chapter with a brief account of "not racism," because, as I will argue, "not racism" has come to play an important role in structuring the discourse of climate change and migration. This chapter examines how the logic of white exoneration associated with "not racism" finds expression in a recurring methodological motif found repeatedly throughout the discourse: the explicit repudiation of climatic determinism and the commensurate turn to "complexity" as the preferred ontology for characterising the relationship between climate change and migration. We might label this motif an example of what Gurminder Bhambra calls "methodological whiteness."[7] I argue that the explicit and regular rejection of determinist reasoning, examples of which are plentiful in climate change and migration discourse, functions to distance those who invoke it from the racisms presupposed or implied in signifiers such as environmental determinism, climatic determinism or, simply, determinism. Much like Lentin's "not racism" and Goldberg's notion of race denialism, the refusal of climatic determinism is a declarative statement in which the speaker is effectively saying that to construct climate change as a problem of migration is "not racism." In this way, by narrowly associating racism with environmental determinism, the determinist refusal allows those who invoke it to set themselves apart from the racisms implied by the signifier "determinism," thereby creating the powerful illusion that if determinist reasoning is racist, those who refuse it are not.

This chapter marks an important transition in the structure of my overall argument about racial futurism. In chapter 2, I sought to show how the figure of the climate migrant/refugee is racialised, while in chapter 3, I examined how the international discourse on climate change and migration is severed from any meaningful relation to the history of racism. My concern in this chapter is to trace how the discourse on climate change and migration gains distance from its racial connotations, those, for example, traced in the preceding chapters. Chapters 5 and 6 address more directly the *futurism* of racial futurism. But to properly contextualise those later chapters, the present chapter examines in detail how the repeated repudiation of determinist reasoning in the discourse on climate change and migration mimics the logic of "not racism" through a methodological appeal to notions of complexity and futurity. For some readers, zeroing in on what might appear to be a trivial aspect of the discourse may seem a little pedantic, maybe even a distraction from the urgency of climatic injustice. My contention, however, is that doing so is necessary to diagnose racial futurism properly. This is because at stake in the refusal of determinist reasoning are urgent issues about race and racism that ought to figure centrally in discussions about migration on a warming planet, but which, for reasons we will soon see, are easily overlooked. Most notable among these are important considerations concerning the power to define racism. In fact, for Lentin, "not racism" is more than simply the denial of racism. Its greater significance lies in the way it "entails the constant redefinition of racism to suit white agendas," and, thus, how it "goes to the heart of the question of who gets to define what racism is."[8] Indeed, for Lentin, one of the main fronts in anti-racist struggle concerns the very definition of racism. And this matters because, while seeing racism in the claim that climate change could result in mass migration to Europe is easy, deciphering its, apparently, more benign expressions is far more challenging without some understanding of the history of race and racism.

To illustrate this, Lentin gives the example of political scientist Eric Kauffman who insists that the desire to curtail immigration on the basis of what he calls "racial self-interest" is not racist. By arguing that racially motivated immigration control is not racism, Kauffman reduces whiteness to an identity no different than other forms of racialised identity. Yet, in taking this position, Kauffman overlooks not only the historical fact that immigration control has always been a structural feature of state racism[9] but also the fact that race, by any honest reading of its history, is first and foremost a "technology of power" not an identity.[10] From the perspective of contemporary antiracism, Kauffman's position is one of the hallmarks of contemporary white power. By redefining the terms on which race and racism are conceptualised, he seeks to shore up the structural power of whiteness. "Not racism" is a logic central to this tactical redefinition of racism.

This chapter is pivotal to my overall argument because, as we will see, the refusal of climatic determinism does something similar within the discourse on climate change and migration. It defines racism in a manner suitable to the needs of racial futurism. By defining climatic determinism as an unfortunate, outmoded world view, the refusal of climatic determinism implies that both climatic determinism and its underlying racism are irrelevant to understanding *contemporary* discussions of climate change and migration. That it does so on methodological grounds only clarifies its methodological whiteness. Part of what I want to put forward in this chapter is that popular and scholarly debate about climate change and migration is today badly served by the refusal of climate determinism inasmuch as it shields the debate from confronting a more nuanced, historically grounded, structural understanding of racism.

One of the challenges in making this argument concerns the fact that the discourse on climate change and migration has been unfolding in a world-political context structured by the *postracial*. I explain the postracial in more detail later, though for the purpose of immediate clarification, I use it to describe the sociopolitical condition in which racism continues to be a dominant feature of Western political culture even as it is routinely denied.[11] For our purposes, this means that because race has no explicit visibility in the discourse, aside from overstated fears about South to North mass migration (although from the example above, we can see how even this racism can be dismissed as "not racism"), the repeated rejection of environmental determinism, which I take to be an expression of "not racism," only really operates as such by implication—if environmental determinism implies racism, the rejection of environmental determinism implies "not racism." The challenge here is compounded by the overwhelming tendency in Western political culture to define racism as individual or institutional expressions of hatred and/or violence directed towards a racial other, usually anchored by some biological/natural referent (e.g., genes, blood, environment). In this way, racism is usually understood to be an aberration from the otherwise open, tolerant attitude often said to define liberal democratic societies, something one would expect of the "deplorables" but certainly not something one would expect of enlightened liberalism. But such a narrow interpretation of racism is unhelpful because it neglects to consider how racism is a structural feature of Western political culture, not simply an aberration.[12]

Later in the chapter I account for the historical origins of not racism as a signifier of racism. There I explain how Lentin arrives at her conclusion that "not racism" consolidates whiteness in the context of the postracial, analogous, for example, to what Eve Tuck and Wayne Yang have called a "move to innocence"[13] and what Robyn Wiegman would call the disaffiliation of white liberalism from white supremacy.[14] At stake here is the idea that the structural power of whiteness has been historically reconfigured since the early 1980s, as both an identity *and* political subjectivity, in such a way that its structural power is, paradoxically, reinforced even as it tries to distance itself from the idea that racism is structural. This historical context is, therefore, of vital importance, as it clarifies how the rejection of environmental determinism as an instance of not racism reframes the discourse on climate change and migration as an expression of whiteness, albeit one now expressed in the language of complexity.

Further, by accounting for the historical production of the logic of not racism, we might start to understand how the discourse on climate change and migration can be understood today as a kind of revival of environmental determinism, an example of what some have called *neo*-environmental determinism.[15] The latter has been described as a form of future-oriented environmental determinism, whose legitimacy I would hasten to add, rests on its denial as such. In this way, neo-environmental determinism mimics the operation of not racism—it is predicated on race denialism even as it reinforces whiteness. Thus, neo-environmental determinism, in the form found in contemporary climate-migration discourse, is consistent with Robyn Wiegman's important insight that "the hegemonic formation of white identity today must be understood as taking shape in the rhetorical, if not always political, register of disaffiliation from white supremacist practices and discourses."[16] Only by grasping this peculiar formation can we begin to grapple with the way race and racism are conceptually and politically reproduced in the wider political context of climate migration discourse and practice.

## THE REJECTION OF ENVIRONMENTAL DETERMINISM

What is environmental determinism? And what is the rejection of environmental determinism? Environmental determinism (also referred to as climatic or geographical determinism) is a discredited pseudo-science associated with scientific racism which gained notoriety from the late nineteenth century to roughly the mid-twentieth century. Its principal claim is that social and political phenomena can, in the last instance, be attributed to environment, climate or geography. A full genealogy and critique of environmental determinism lies beyond the scope of the present chapter. For our purposes, environmental determinism is most closely associated with Friedrich Ratzel, whose concept of *Lebensraum* ("living space") would go on to inform Nazi territorial strategy, as well as Ellen Churchill Semple, Ellsworth Huntington, Griffith Taylor and Halford McKinder, all of whom would play a key role in shaping the academic discipline of geography.[17] Environmental determinism would later come under heavy criticism by Carl Sauer and the Berkeley School and is nowadays, more or less, remembered as a form of scientific racism that served to legitimise European and American imperialism in the first half of the twentieth century.[18] Environmental determinism has, however, undergone something of a revival in the past few decades and is now associated with writers such as Jeffrey Sachs, Robert Kaplan and Jared Diamond, whose popular book *Guns, Germs, and Steel* earned a Pulitzer Prize in 1998.[19] A full critique of the structure and legacy of the racisms advanced by environmental determinism is yet to be written.[20]

The rejection of environmental determinism and the corresponding turn to complexity are at the heart of the epistemological disjuncture of climate migration discourse discussed in previous chapters. It is implied, for example, every time a direct causal link between climate change and migration is dismissed on methodological grounds in favour of an approach in which climate migration is said to be a *complex*, multicausal phenomenon. The denunciation of determinism is furthermore implied when claims about mass migration in relation to climate change are written off as alarmist, or when geopolitical narratives about transboundary migration leading to political violence and insecurity are said to be either methodologically flawed or excessively militarist. It is noteworthy, however, that even as security discourses about climate change and migration are routinely criticised for being determinist, the U.S. military also appears to embrace the rejection of determinist reasoning in favour of complexity, a point I return to later and in chapter 5.[21] In fact, as we saw in chapter 1, the dismissal of causal reasoning is a standard disclaimer found throughout the discourse. It is marked, however, by a fundamental slippage, which is that climate migration cannot be considered *climate* migration, unless climate change is elevated as the primary, even if not the exclusive, determinant of migration. It is the racio-political stakes contained in this slippage that interest me in this chapter.

What exactly do I mean by the rejection of environmental determinism? What form does it take? How and with what consequences does it relate to race and racism? To answer these questions let us first consider some prominent examples taken from the social science and policy literatures on climate change and migration. The first of these comes from a highly influential report published in 2011 by the U.K. Government Office for Science called *Migration and Global Environmental Change* (hereafter *The Foresight Report*).[22] Soon after it was published *The Foresight Report* gained prominence within the fields of migration, development and humanitarianism for the way it sought to redefine the relationship between climate change and migration in the language of human security, humanitarianism and adaptation. It did so, moreover, at a time when climate security and militarism had come to dominate international discussions about migration and climate change.[23] Whereas the broad discourse on climate security typically constructs the climate migrant/refugee as a catalyst for political violence, *The Foresight Report*, by contrast, constructs the relation between climate change and migration in more anodyne language. It argues that migration should be viewed as a legitimate adaptive response to climate change rather than a failure to adapt, which is implied when the figure is said to be a security threat or pending humanitarian crisis.

*The Foresight Report* is additionally important because it cautioned *against* the assumption that migrants respond to environmental change by moving away from risk. Instead, one of its lasting insights has been that many inadvertently migrate *into* spaces of even greater risk where they become trapped and no longer able to migrate. These so-called trapped populations, it argued, represented a point on the migration continuum easily overlooked when climate change is said to result in mass transboundary migration.[24] The fact that the arguments appearing in *The Foresight Report* would feature prominently in international discussions on climate change adaptation, development and migration policy in the period leading up to and in the wake of the Paris Agreement is good reason to take them seriously. To give just one example, one of the key policy pillars adopted by the World Bank draws its reasoning from *The Foresight Report*, namely the idea that, if properly managed, migration can be an appropriate even desirable adaptive response to climate change. These arguments have also prompted significant critical attention as examples of neoliberal governance and biopolitical rule,[25] a point I return to in detail in chapter 6.

For our immediate purpose, though, we can simply locate *The Foresight Report*'s rejection of environmental determinism in the following statement:

> There are a number of existing estimates of the "numbers of environmental/climate migrants," yet this report argues that these estimates are methodologically unsound, as migration is a multi-causal phenomenon and it is problematic to assign a proportion of the actual or predicted number of migrants as moving as a direct result of environmental change. A *deterministic approach* that assumes that all or a proportion of people living in an "at-risk" zone in a low-income country will migrate neglects the pivotal role that humans take in dealing with environmental change, also ignores other constraining factors which influence migration outcomes.
>
> This is not to say that the interaction of migration and global environmental change is not important: global environmental change does have real impacts on migration, but in more *complex* ways than previous cause-effect hypotheses have indicated.[26]

Two considerations are noteworthy here. First, the refusal of determinism in *The Foresight Report* is based on methodological grounds. This is important because it echoes the first generation of critiques levelled against environmental determinism in the mid-1920s, when determinism was challenged on *methodological* as opposed to ideological grounds.[27] According to William Meyer and Dylan Guss, it was not that environmental determinism offered a racist justification for European imperialism that troubled its early critics. It was that the environment inadequately explained social relations. It was, according to these early critiques, simply bad science.[28] *The Foresight Report* repeats this same logic, particularly when it claims that a direct causal relation between climate change and migration is "methodologically unsound." But the second consideration is that, even while *The Foresight Report* refuses determinism on account of the way determinism is said to link a direct environmental cause (climate change) with an immediate social effect (migration), the report does not rule out an *indirect* relationship between cause (climate change) and effect (migration). In fact, another key claim of *The Foresight Report* is that while global environmental change will almost certainly affect migration, it will only do so by affecting a range of *intermediary* drivers of migration, such as agricultural conditions, labour markets, political conditions, health and so forth (see Figure 4.1). Importantly, *The Foresight Report* characterises this indirect relation of causality as "complex" as opposed to deterministic. What the report says is that the effect of climate change on the decision to migrate is mediated by these intermediary variables (e.g., in Figure 4.1, these are political, social, economic, demographic and environmental), most of which are hard to predict and, thus, relationally complex. Later, however, we will see how this logic of indirect causality closely resembles Ellsworth Huntington's description of "environmental migration," which, in turn, will allow us to see how environmental determinism seems to reappear in *The Foresight Report* in the guise of its opposite.

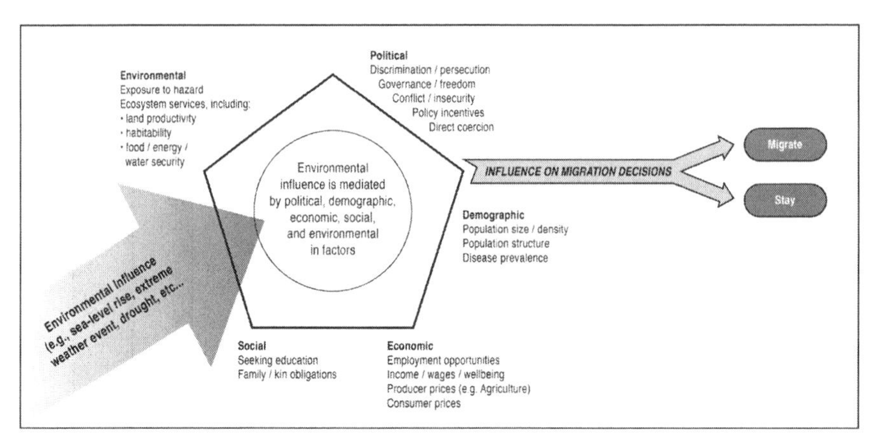


**Figure 4.1 The indirect influence of environmental change on migration decisions. Source: Image adapted from U.K. Foresight, *Migration and Global Environmental Change* (London: U.K. Office for Science, 2011).**

The second example comes from a more obscure but no less important international policy text, *Addressing Climate Change and Migration in Asia and the Pacific*, published by the Asian Development Bank in 2012. The conceptual innovation of this text is that it connects up the figure of climate migrant/refugee to questions of global finance, risk and uncertainty by setting out the rationale for using parametric metric insurance to cover the costs of "climate-induced migration." What makes this report significant is that it openly speculates on how climate-induced migration might be financialised, thus paving the way for treating the risk of migration as a fungible commodity. Like *The Foresight Report*, the Asian Development Bank report also rejects determinist reasoning. It says, for example, that "[t]he relationship between climate change and migration flows is often thought to be of a deterministic nature. . . . Many empirical studies show, however, that this relationship is far more complex, and is compounded by a wide range of social, economic, and political factors."[29] I discuss this text in more detail in chapter 6.

The rejection of environmental determinism is found not only in international policy texts. The same pattern also appears throughout the scholarly literature, a clear example of which is found in Robert McLeman's *Climate and Human Migration: Past Experiences, Future Challenges*. McLeman argues, for example, that

> to avoid lapsing into a new paradigm of "climatic determinism" that focuses on the role of climate to the exclusion of other influences on migration, it is important to knit the theoretically grounded knowledge of migration behaviour

> that has been developed over long periods of time by its specialists with the empirical evidence of climate-related migration that has been generated in recent decades through research in other fields.[30]

Here, McLeman clearly distances himself from the legacy of climatic determinism and proposes in its place a form of empiricism that acknowledges the multicausal nature of migration and which squares migration behavioural theory with actual events of what he calls "climate-related" migration. Just like *The Foresight Report*, McLeman refuses determinism on methodological grounds, but adds that, for him, determinism's flaw consists in the fact that it excludes "other influences on migration."[31] Another example of the determinist refusal comes from Etienne Piguet whose argument, examined in the previous chapter, on the disappearance and subsequent reappearance of the environment in migration studies, includes the following: "I want to look at [the relation between environment and migration] not in a deterministic way, as a singular and external driver of the movement of people, but rather as the *complex* product of interactions between nature and society."[32] Again, like *The Foresight Report* and McLeman, Piguet explicitly refuses deterministic reasoning in favour of an approach which acknowledges the inherent complexity of the relation between climate change and migration. And a final example can be found in work by François Gemenne, whose research, like that of McLeman and Piguet, has also been highly influential in shaping the field of environmental migration studies. The title of his 2013 paper captures precisely the determinist refusal: "Migration Doesn't Have to Be a Failure to Adapt: An Escape from Environmental Determinism."[33] In it, Gemenne argues that quantitative predictions of the number of people who will be displaced by climate change "reflect a deterministic perspective." The idea that the so-called climate migrants can be quantified, he argues, "is *deeply rooted* in environmental determinism, which assumes that people's course of action is exclusively determined by their environment."[34]

This is only a small sample of texts that deploy the determinist refusal. One could easily reference numerous others.[35] I highlight these five, though, because their authors are all-important figures within the epistemic community now defining the terms of the international discourse on climate change and migration.[36] All have played a central role, for example, in institutionalising the claim that the effects of environmental change on migration are complex and multicausal and that migration should be considered a legitimate adaptive response to climate change as opposed to a failure to adapt. It is noteworthy that while important substantive, disciplinary and argumentative differences exist across all five interventions, they share in common a clear attempt to gain distance from environmental determinism on methodological as opposed to ideological grounds. It is, therefore, on this basis that we can say that the refusal of determinism has become foundational to the current research and policy debate on climate change and migration. While it is certainly true to say that determinism continues to hold considerable sway in popular media and political representation,[37] we can also safely assume that determinism, at least when conceptualised as direct causality (e.g., climate change causes migration) or alarmism (e.g., the barbarians at the gates), appears to have minimal influence nowadays in international policymaking circles.

## THE EFFECTS OF THE REFUSAL OF ENVIRONMENTAL DETERMINISM

Taken together, these prominent examples illustrate how the explicit and repeated rejection environmental determinism appears to be a structural feature of the science and policy literature on climate change and migration. But how might we interpret this rejection of environmental determinism? What does it do to the meaning of the discourse? What are its effects? Although much can be said here, I wish to highlight four such effects.

First and foremost, the refusal of determinism is a claim to truth. Each of the five examples cited earlier appeals to "complexity" in order to explain migration in the context of climate change. When the relationship between climate change and migration is said to be a complex, multicausal relationship or when migration is said to be the complex outcome of social and environmental forces, the apparent truthfulness of these statements is set *against* the specious method of determinist reasoning. Thus, the first effect of the determinist refusal is that, in gaining distance from determinist thought (a falsehood), we are moved away from the pseudo-science of "environmental determinism" and closer to the "truth" of the relation. According to this logic, taking account of the complex nature of the interrelationship between climate change and migration would give a more reliable, "truthful" account as to why people migrate than would a clunky, outmoded determinism.

That the turn to complexity is a claim to truth may sound reasonable enough, but it is connected to a second, related effect: the determinist refusal also implies and, therefore, establishes what it assumes is the *truth* about racism. To the extent the contemporary discourse on climate change and migration involves the explicit rejection of environmental determinism and, thus, racism, it has the effect of defining racism in a very precise way. In keeping with the long-standing liberal definition of racism, it defines racism as "a moral wrong based on bad science."[38] In this way, it associates racism with individual ignorance, immorality and dubious science, as opposed to conceptualising racism as a *structural feature* of modernity. It assumes that racism is an aberration to the normal workings of modernity, as opposed to one of its founding conditions.

This is, of course, not immediately obvious when we take the earlier examples at face value. In fact, none make any *explicit* reference to a truth about racism. When taken on their own terms, they are simply methodological statements. We can, however, recover this racial meaning by turning to the historical origins of racism. According to Ladelle McWhorter, the concept of racism only emerged in the West in the mid-1930s when it was first used to denounce the racial science that underpinned Nazism.[39] At the time, racism was not conceptualised as a structural feature of Western modernity. It was instead simply the name given to a "false theory" of race, one that would connote the "denial of science" and the "wilful misrepresentation of scientific knowledge."[40] The result was that racism would henceforth come to describe an irrational or ignorant world view, while the term racist would be applied to those who were said to suffer from a kind of cognitive or moral failing, owing to their inability or reluctance to accept that scientific racism was nonsense. This was the typical liberal understanding of racism. However, as Lentin goes on to clarify, this historical version of anti-racism was also only concerned with refusing racial differences *among Europeans*, yet remained perfectly content to accept that race could be used to explain whatever differences were said to exist between Europeans and non-Europeans.[41] Indeed, even as anti-racism first took shape during this period across the West, the global war on people of colour continued unabated. It was not as though the vilification of Nazi racial science resulted in the end of racism. The global colour line that underpinned nineteenth-century European imperialism would persist long after World War II. And even then it would simply get rearticulated in the language of culture rather than eugenics and biological difference, which, up until the 1940s, were its predominant modes of expression and not just in Nazi Germany but across the West.

Stepping back from this brief history of racism, my point here is that a similar construction of racism is at stake in the determinist refusal. Even as it rejects environmental determinism—a pseudo-science used to support a specific nineteenth-century racist world view—the determinist refusal, like Lentin's "not racism," is not the denunciation of racism *tout court* so much as the denunciation of a specific understanding of racism, one that constructs racism as a methodologically flawed world view. In this way, the determinist refusal defines racism in very narrow terms, while ensuring that structural racisms pass unnoticed. Thus, rather than conceptualising racism as an endemic feature of the modern world, the determinist refusal has the peculiar effect of leaving structural racism unchallenged even while appearing to do the opposite.

A third effect of the determinist refusal is that more than simply a statement of methodological preference, it is also an ontological claim. It holds that the relation between climate change and migration is not deterministic, linear and causal, but complex, non-linear, emergent and unpredictable. Framing the relationship between climate change and migration as complex, in turn, has important consequences on how governing migration in the context of climate change is imagined. For example, from the perspective of military planning and capitalist political economy, climate change is increasingly grasped as a slowly unfolding complex risk event, while migration is imagined as the symbolic manifestation of this complex risk. Accordingly, any migration that might result from climate change cannot be fully known in advance, which means that governing climate-related migration entails governing the *threat of migration*, as opposed to migrants themselves. For now, though, we can say that if *cause* and *effect* are said to comprise the basic ontology of determinist thought, then from the perspective of the determinist refusal, *complexity*, *non-linearity* and *emergence* comprise the basic ontology for defining the relationship between climate change and migration. Later in the chapter, I argue that this ontology forms the basic building blocks of neo-environmental determinism, a logic which, according to Mike Hulme, involves reducing the future to climate change. As we will see in chapter 6, the ideas of potential and risk are actually quite fundamental to the way racial futurism operates.

Finally, and, perhaps most consequential, the determinist refusal is also a claim to novelty. Environmental determinism is assumed to be an archaic or crude form of thought and, thus, like the racism it justifies, one that can and should be safely relegated to the past. In contrast, the idea that the relation between climate change and migration is complex is given the appearance of novelty. Complexity is said to supersede determinism as a superior way of knowing this relation. It promises to modernise how we understand the relationship between environment and migration. No longer encumbered by outmoded assumptions of linear cause and effect, the turn to complexity imagines a new frontier of research into the relationship between climate change and migration, one that will result in a more sophisticated, truthful knowledge that will, ultimately, result in a more informed policy for governing the impending crisis.

It is important to understand, however, that there is nothing remotely novel about the claim that migration is multicausal and irreducible to environmental change. Not only has this argument been made repeatedly in contemporary policy and research, the determinist refusal and the commensurate turn to complex explanation has been a mainstay of academic geography since the late nineteenth century. Indeed, as the English climatologist Gordon Manley wrote in 1944, "One of the most noteworthy developments *during the last sixty years* has been the steady reaction, indeed objection, on the part of many geographers to the simpler determinism." "Geographers and historians alike," he goes on to tell his reader, "have pointed out the infinite complexity of the relationships between man and his surroundings; and have emphasized the need for detailed study in the field rather than categorical generalizations smelling of the lamp."[42] It is hard to imagine a better way of contextualising the determinist refusal than this.

Together, these four effects of the refusal of environmental determinism give the impression that the science and policy discourse on climate change and migration bears no historical relation to environmental determinism. Instead, they generate a sense that in our contemporary world, aided now by developments in science, methodology and a critical perspective on racism, a more sophisticated, informed understanding of the relationship between climate change and migration is emerging that transcends the ideological work and legacy of environmental determinism. It is noteworthy, however, that even as *The Foresight Report*, McLeman, Piguet and Gemenne distance themselves from environmental determinism, they are all nevertheless of the view that climate change should be afforded some differential, *primary* status in order to understand the unique contribution it makes to migration. McLeman, for example, uses the notion of climate-related migration, even while insisting that climate change is never the exclusive determinant of migration. Gemenne does the same, when he refers to "climate-induced migration." Piguet is slightly more nuanced, allowing at least some space for conceptualising climate change as a specific crisis of capitalist political economy. His ultimate aim however, is to foster a "pragmatic" approach which conceptualises migration as "the product of complex interactions between nature and society."[43] In each case, "climate" is given elevated, primary status to explain migration. Thus, for each of these authors, while the historical composition of social and political life absolutely matters when explaining social outcomes like migration, they all seem to be suggesting that, in the final instance, climate matters more.

These examples should, therefore, give us a pause to reconsider how we read the determinist refusal in the contemporary debate on climate change and migration. Is the refusal of determinism and the turn to complexity really that different from the environmental determinism that did so much to naturalise European imperialism in the latter years of the nineteenth century and first half of the twentieth? Part of the difficulty in answering this question is that the claims these authors are making are predicated on rather vague, unspecified readings of determinism. Determinism clearly functions for each of these authors as a problematic approach. But never do they give any detail about what determinism actually refers to beyond the simple idea that climate "determines" or "causes" migration, or that the relation is linear or binary, or that climate change will result in mass migration.[44] When determinism is rejected, it is often simply assumed that the reader knows in advance what determinism is and why it has been discredited. Nor do these authors ever really specify who precisely adopts the deterministic approach they worry about. At best, deterministic reasoning is said to underpin rudimentary quantitative speculation of the kind we saw in chapters 1 and 2. Such speculation is often attributed to sensationalist news media, or very specifically to the environmental scientist Norman Myers whose quantitative speculations (e.g., 200 million climate refugees by 2050) lend so well to alarmist media representations of climate change and migration. Determinism is also often said to sit at the heart of climate security narratives. Gemenne, for example, draws a series of associations which equate alarmism, securitisation and quantification which are said to be underpinned by determinism.[45] Elsewhere he and Olivia Dun distinguish between sceptics and alarmists, where the former are said to be sceptical of a direct causal link between climate change and migration, whereas "reports linking climate change with security issues usually side with alarmists."[46] In fact, the argument linking climate change and insecurity is itself routinely said to be founded on determinist assumptions that neglect the complex empirical dimensions of political economy.[47] Yet, all too often, those who denounce determinism rarely, if ever, give a comprehensive account of what they mean by determinism or of how their approach differs from the environmental determinisms of the late nineteenth and early twentieth centuries. And this matters because, without a clear understanding of what these authors actually mean by determinism, their rejection of determinist thought does far more to obscure the similarities between their use of complexity and determinism than it does to clarify their differences.

Two concrete examples illustrate my point. As prominent texts on the geopolitics of climate change, migration and security, both are supposedly underpinned by determinist assumptions. However, on closer inspection, this doesn't appear to be the case. The first is Jared Scott's popular climate security documentary film *The Age of Consequences*. Named after an influential report published jointly in 2007 by the U.S.-based Center for Strategic and International Studies and the Center for a New American Security (discussed in chapter 5), *The Age of Consequences* premiered at the Toronto HotDocs film festival in 2016 and was later nominated for a News and Documentary Emmy award. Given both its notoriety and its connection to U.S. military planning, we can assume that *The Age of Consequences* is a reasonable, even if polished, account of current military thinking on climate change. Although much can be said about the film, it is enough simply to point out that the film could not be clearer that the U.S. military conceptualises the relationship between climate change, security and migration within a non-linear and complex ontology. It tells us, for example, that climatic events, even minor ones, in one part of the world, could yield violent consequences with cascading effects, including migration, in places far removed from the initial site of environmental turbulence. Moreover, not unlike *The Foresight Report*, *The Age of Consequences* is clear that the path between cause and effect is mediated by all manner of intervening variables, which makes causal attribution impossible. The report published a decade earlier argues the same. It states quite explicitly that "climate change is a manifestation of phenomena that are complex in the technical sense of that word" and, later, that "*nonlinear climate change will produce nonlinear political events.*"[48] It is also true that both the film and the report are themselves firmly alarmist in both tone and substance. To suggest, then, as Gemenne does, that a "deterministic perspective is a common feature of works written from an alarmist viewpoint" or to argue that narratives of geopolitical insecurity, determinism and alarmism are of a piece is actually quite misleading.[49]

The second example belongs to Colin Kelley, Shahrzad Mohtadi, Mark Cane, Richard Seager and Yochanan Kushnir whose influential study of the supposed environmental underpinnings of the Syrian civil war is often held up as definitive proof that climate change will result in migration, violence and protracted war. It is, for example, cited explicitly in Scott's film. Here, again, much can be said, but for our purposes it is noteworthy that the study makes no claim about a causal, linear relationship between climate change, migration and the devastating political violence wrought by the Assad regime on the people of Syria. It certainly maps climate change onto the pattern of migration that preceded the revolution, but it is also very clear that the interrelationship between migration, drought and the revolution was as much a function of a complicated set of social and political relations as environmental change:

> Civil unrest can never be said to have a simple or unique cause. The Syrian conflict, now civil war, is no exception. Still, in a recent interview, a displaced Syrian farmer was asked if this was about the drought, and she replied, "Of course. The *drought and unemployment* were important in pushing people toward revolution. When the drought happened, we could handle it for two years, and then we said, 'It's enough.'" This recent drought was likely made worse by human-induced climate change, and such persistent, deep droughts are projected to become more commonplace in a warming world.[50]

What these two examples point towards, then, is that the novel ontological claim implied in rejection of environmental determinism and the turn to complex reasoning for conceptualising the relationship between climate change and migration betrays a misreading of determinism as brute causation, analogous to billiard balls on a table. Yet, when we look at actual examples of the kinds of reasoning that are said to be deterministic, we find the opposite, forms of reasoning underpinned by an ontology of complex systems.

We can track the slippage in this logic even further. In three of the four examples noted earlier—*The Foresight Report*, McLeman and Gemenne—the determinist refusal is used to set up the claim that migration should or, at least, can be considered a legitimate adaptive response to climate change as opposed to a failure to adapt. On these accounts, migration is said to be a naturally occurring adaptive strategy that humans have always used to respond to environmental change. Migration is thus said to be a *natural* aspect of human life and, as such, one that should be considered part of the solution to climate change not the main problem. In this way, these accounts associate determinist reasoning with adaptive failure. Yet, when we turn to some of the main texts on environmental determinism written in the early twentieth century, the picture appears to be much more complicated. Instead, what we find is a literature centrally preoccupied with adaptive migration. In the paleo-evolutionary accounts of environmental determinists like Ellen Churchill Semple, Ellsworth Huntington and Griffith Taylor, the geographical distribution of the so-called races—what Gilroy would call the racial nomos of Western racism[51]—was said to be a function of their adaptive capacity. Take Huntington as just one prominent example:

> To sum up the whole hypothesis of the relation of climate to civilisation, here are the factors as I see them at present. Most parts of the world are so well populated that any adverse economic change tends to cause distress, disease, and a high death rate; migration ensues among the more energetic and adventurous people. Perhaps the most common cause of economic distress is variations in weather or climate which lead to political disturbances and this again is a potent cause of migrations. The people who migrate perforce expose themselves to hardships and their numbers diminish until only a selected group of unusually high quality remains.[52]

Two points are worth making here. First, Huntington's entire determinist hypothesis bears a striking resemblance to what is nowadays said to be the complex relationship between climate change and migration. Like *The Foresight Report* (Figure 4.1), the pattern of causation in Huntington's reasoning is not direct but indirect. Environmental change is a driver of migration, but only to the extent it is mediated by local political economy. Second, what Huntington presents is precisely the logic of adaptive migration. Groups of people—so-called races—were said to have adapted to the environments in which they live, while it was migration that enabled them to adapt to their changing circumstances. Huntington was a racist and, therefore, sought to explain people, indeed the so-called races, on the basis of nature or, in his case, environment, although for him it seems this explanation also involved a complex mixture of migration, environment, economy and adaptation. By this reasoning the "natural" spatial order of race—racial nomos—owed everything to adaptive migration. But to assume, as the determinist refusal so often does, that determinism equates to alarmism, or that determinism equates to the failure to adapt, or that it is inherently securitising, betrays a fundamental misunderstanding of environmental determinism. And this is crucial because when writers interpret environmental determinism in such narrow terms, they obscure the fact that adaptive migration was in fact central to the logics of environmental determinism not its antithesis.

But what does any of the preceding discussion have to do with race and racism? How does it clarify Lentin's conception of "not racism"? The main point I wish to draw from the preceding discussion is that by disaffiliating the contemporary discourse on climate change and migration from environmental determinism, the determinist refusal effectively decouples the discourse from any meaningful connections that might be drawn between it and the history of race and racism. That is, by obscuring what is meant by environmental determinism and by drawing a series of misleading associations between environmental determinism, alarmism and insecurity driven by climate change, those who invoke the determinist refusal are made to appear as though their construction of the relation between climate change and migration as a "complex" phenomenon transcends race and racism. Which brings us to the very crux of the argument about "not racism" developed in this chapter: the primary effect of the determinist refusal is that it defines racism and colonialism as artefacts of the distant past which have no immediate relevance to the contemporary debate about climate change and migration. Environmental determinism is, in this sense, construed as part of the ideological edifice of European imperialism and colonialism, both of which belong to a bygone era, such that the refusal of determinist reasoning amounts to the refusal of the racial reasoning that underwrote both. The trouble with this argument, however, is that it obscures how race, racism and colonialism *continue* to shape the contemporary world order, including the discourse on climate change and migration. In this way, the determinist refusal is powerful precisely because it lends the field of climate change and migration studies an air of innocence by making it appear as though the field has transcended race and racism. Yet, as we saw in the preceding two chapters, this is clearly not the case.

## WHITENESS, "NOT RACISM" AND THE POSTRACIAL

The denunciation of determinism is a powerful motif of white exoneration. Which is to say, if environmental determinism signifies racism, then its refusal can be understood to signify the opposite: not racism.[53] The rejection of environmental determinism also mirrors the wider sociopolitical condition of postraciality. The postracial is a controversial concept in the sociology of racism. For some, it expresses the view that race and racism are no longer relevant for explaining political and social life. According to this interpretation, the postracial condition is one in which the West has successfully managed to overcome the historical legacies of racism. In America, for example, although historical phenomena such as slavery and segregation are undeniably a part of America's racist past, those who endorse this postracial perspective will often point to the figure of Barack Obama as evidence that America is no longer a racist state. By this reasoning, if America was racist, Obama would never have become president.

David Theo Goldberg provides, however, a much-needed counter to the rather bizarre claim that the West, in general, and America, in particular, has somehow transcended racism. His argument is that the condition of postraciality is, in fact, the condition of racism without racism.[54] What Goldberg means is that the postracial is not what comes after racism, as though racism might be neatly confined to the past, as though race no longer matters as a category for contemporary social or political analysis. Rather, for Goldberg, the postracial is the very continuity of racism, albeit in the absence and often explicit denial of racial reference. It designates the way in which racisms continue to organise experience unevenly in the present without any explicit reference to race or racism. Thus, for Goldberg the "postracial is the sociality of heightened raciality in the name of its denial."[55] The postracial is precisely, then, the condition in which resurgent white nationalisms flourish nowadays through the vigorous disavowal of their racisms, or the condition that allows Donald Trump to insist that he was not being racist after "accusing six Congress members, including Alexandria Ocasio-Cortez, Ayanna Pressley, Rashida Tlaib, and Ilhan Omar, all of whom are women of colour, of being 'savages.'"[56] But it is also the condition in which liberalism, as well as some forms of left politics, actively forget *their own* racial histories.

But even while all these various racisms may seem like accidents of history, aberrations to otherwise normal political life in the West, it is important to acknowledge that the postracial condition is no accident. Like all political conditions, it is one that has been actively fostered. Following Ann Stoler, we can say that the postracial is akin to the political condition she calls "colonial aphasia":

> Aphasia is a condition in which the occlusion of knowledge is at once a dismembering of words from the objects to which they refer, a difficulty retrieving both the semantic and lexical components of vocabularies, a loss of access that may verge on active dissociation, a difficulty comprehending what is seen

> and spoken. Colonial aphasia as conceived here is a *political* condition whose genealogy is embedded in the space that has allowed Marine Le Pen and her broad constituency to move from the margin and extreme—where her father was banished—to a normalized presence in contemporary France.[57]

For Stoler, colonial aphasia is not only the active disremembering of colonialism. Nor is it just the cognitive inability to retrieve historical meaning from certain words or phrases or the act of dissociating oneself from the colonial histories and legacies that constitute white Eurocentrism. Like the condition of postraciality, it is a *political* condition, fostered to suit a specific set of interests. It is also, as Stoler notes, precisely the ground in which racist social projects take root. This then is one of the obvious dangers of the postracial condition. It must be actively fostered for the onto-epistemological structure of whiteness to endure, which is especially worrisome when we consider that the political discourse of climate change is primarily one about the future. In its rush to ward off future climatic disaster, a concern for climate change contains within it the tendency to disremember the past, including historical formations of race and racism. This is partly why certain forms of climate justice are so important. They preserve the memory of colonialism within the discursive context of climate change science and policymaking.

Alana Lentin has usefully extended Goldberg's conception of the postracial, arguing that the declaration of "not racism"—the act of refusing that a particular action or statement is racist—is nowadays "a central formulation for the expression and legitimation of racism."[58] But where does this peculiar condition come from? How can we account for the fact that racism is now regularly expressed in Western political culture in the language of its opposite? Lentin's answer to these questions draws partly from the field of moral philosophy where racism is defined mainly in terms of individual moral failure, where it is said to be an immoral act, wielded with the intention of causing harm. For Lentin, however, when racism is wrongly conceived solely in terms of individual moral failing, several things happen. One is that people, especially white people, oftentimes react strongly when made to confront their implication in racism. The automatic assumption is that their individual moral character is being called into question, that they themselves are bad people. Another, and this is vastly more important for Lentin, is that we lose sight of the fact that race is foremost a system of structural power. Or as she puts it, a "technology for the management of human difference, the main goal of which is the production, reproduction, and maintenance of white supremacy on both a local and planetary scale."[59] The result is that, rather than contesting racism as a system of oppression that intersects with "other structures of power—capitalism, gender, sexuality, class and ability," racism is reduced to an attitude that one either has or does not have.[60] It is hardly surprising, then, that "not racism" has emerged today as a hegemonic form of racist expression. No one, least of all actual racists want to be tarred with the moral failings and irrationality of the racist.

But the condition of "not racism" has a history that extends beyond simply moral philosophy. As discussed earlier, Lentin is very clear that "not racism" has its origins in early twentieth-century Europe when racism first came to be defined as a pathological world view associated with the Nazi Holocaust. From this early perspective on racism, a racist was someone who endorsed the pseudo-science of scientific racism. It was also someone who viewed race as a naturally occurring form of difference and, therefore, someone for whom the task of scientific racism was simply to explain what was assumed in advance to be a natural fact. However, as Lentin goes on to explain, two consequences follow from this. One is that racism would thereafter be associated with Nazism and other extreme forms of racial violence like Apartheid and Jim Crow, such that the meaning of racism would eventually come to be fixed by these historical referents. The other is that this fixed conception of racism would henceforth rely on "a narrow interpretation of race as confined to nineteenth-century expressions of eugenicist racial science."[61] The result is that the very concept of racism and its corresponding conception of race that arose during the tumultuous events of the twentieth century would come to anchor contemporary understandings of race and racism and, in doing so, obscure the much longer history of race as a classificatory system, which had informed European racial-colonial rule for the preceding four centuries.[62]

Robyn Wiegman offers a more recent history of not racism. She identifies how from the 1960s, American liberal whiteness was redefined in part by distancing itself from Southern pre–civil rights white racism. The latter had come to express the perceived loss felt by Southern whites following the civil war—the white subject as the subject of injury. Wiegman's primary argument is that such distancing, or in her words disaffiliation, is the foundational manoeuvre that allows all manner of American whites (and not only so-called liberals) to imagine themselves as participating in a postracist political culture that privileges integration over segregation for negotiating the complexities of American race relations. Yet, for Wiegman, liberal whiteness is deeply paradoxical for at the very moment it seeks to be, in effect, antiracist, it disavows its own hegemony. One of the most crucial consequences of this, she contends, is that it allows white Americans to believe they are not racist, while simultaneously, as she puts it, "aggressively solidifying [their] advantage" in American life through, for instance, rollbacks to welfare and affirmative action, and the proliferation of gated communities, now so characteristic of American-style neoliberalism.[63] This is what Wiegman means when she says that "the hegemonic formation of white identity today must be understood as taking shape in the rhetorical, if not always political, register of disaffiliation from white supremacist practices and discourses."[64] But equally characteristic of liberal whiteness, for Wiegman, is the disavowal of the way whiteness is *re*constituted in the very moment it disaffiliates from older, more conventional versions of itself. That is, liberal whiteness disavows its universalism by claiming to be a *particular* form of identity; whiteness particularises from the universal even while it retains for itself the right to define the terms of anti-racism in ways that shore it up. Wiegman, then, goes on to argue that "this split in the white subject—between disaffiliation from white supremacist practices and disavowal of the ongoing reformation of white power and one's benefit from it—is constitutive of contemporary white racial formation."[65] Borrowing from Howard Winant, she refers to this contradictory position as "white racial dualism."[66]

Wiegman's analysis of white racial dualism is indispensable for understanding contemporary discourses on climate change and migration. Not only is this because the structural position of whiteness is recentred in the discourse on climate change and migration through its disaffiliation from determinist reasoning. It is also because of the *other* side of white racial dualism: the way the determinist refusal also entails the "disavowal of the ongoing reformation of white power."[67] The discourse on climate change and migration may well entail the refusal of determinism and, thus, racism. But if Wiegman's history of whiteness and Lentin's history of "not racism" tell us anything, it has to be that when whiteness denounces one form of racism it often regathers itself around another. For Wiegman, liberal whiteness was able to sustain a commitment to ending segregation even while putting in place the political economic conditions necessary to extend and deepen its economic privilege. So, too, Lentin's "not racism" involves the systematic hardening of racism in the name of its opposite. Something similar, I would suggest, is at stake in the discourse on climate change and migration.

## METHODOLOGICAL WHITENESS AND NEO-ENVIRONMENTAL DETERMINISM

The explicit rejection of both determinism and racism is a standard trope in what some now call *neo*-environmental determinism. As Andrew Sluyter has shown, for example, the rejection of environmental determinism and racism forms the basis of Jared Diamond's rearticulation of environmental determinism in *Guns, Germs, and Steel*. William Meyer and Dylan Guss identify a similar pattern in Diamond's book *Collapse*. I point this out simply to identify how the rejection of environmental determinism is a standard disclaimer in neo-environmental determinist thought just as "not racism" is a feature of contemporary racism. If "not racism" finds popular expression in the familiar statement "I am not a racist, but . . . ," then the rejection of environmental determinism expresses a similar logic. It says, "I am not an environmental determinist, but I will explain migration through a primary emphasis on the environment."

What precisely is neo-environmental determinism? How is it exemplified in the contemporary discourse on climate change and migration? According to Meyer and Guss, it is the resurfacing of environmental determinism, albeit in a slightly modified guise. For these writers, environmental determinism can take one of two forms. It is either a form of knowledge that "treats the environment as a separate, simple cause or 'factor' not mediated by culture: something external to culture and influencing it from the outside" or it can be "an overriding or predominant focus on the role of environmental factors relative to the social ones in explaining a given situation of nature-society interaction." The first of these, the *strong* form of environmental determinism, coincides with what some in the field of "environmental migration" oftentimes refer to as a maximalist interpretation of the relation between environmental change and migration.[68] My contention is that the determinist refusal denounces *this* reading of determinism. Thus, the authors of *The Foresight Report* can rightfully claim that their model is not determinist. However, Meyer and Guss's second definition of environmental determinism is slightly more expansive. This minimalist, or *weak*, environmental determinism gives primacy to the environment as an explanation for migration, even while acknowledging that the environment is mediated by social relations. In other words, the outcome of a socioecological interaction, such as migration, may well be multicausal—the complex mix of political, economic, cultural, demographic and environmental variables—but it is the overriding environmental trigger, or final cause, which sets the chain of reactions in motion. This weak version of environmental determinism is clearly evident in the model used in *The Foresight Report* (Figure 4.1). But as we saw earlier, it is also evident in the model used by the U.S. military and by McLeman and Gemenne.

Both strong and weak variants on environmental determinism are evident today. It is, however, the weak form that has become hegemonic in contemporary international climate change and migration discourse. Meyer and Guss identify many variants of weak *neo*-environmental determinism. These are explanations that place "chief or sole emphasis on the environmental elements in a situation at the expense of social ones." Most pertinent to our discussion, however, is that in which *future* social and political life is reduced to, or determined by, climatic or environmental change. Mike Hulme calls this approach climate reductionism, a form of thinking in which "climate is first extracted from the matrix of interdependencies that shape human life within the physical world" and "then elevated to the role of dominant predictor variable."[69]

Winant's conception of white racial dualism is important for understanding how this version of neo-environmental determinism can be understood as a form of methodological whiteness. White racial dualism involves rejecting older forms of racism while reconsolidating whiteness around newly reformulated racisms. The determinist refusal and commensurate turn to complexity to explain migration does exactly this. It reconsolidates whiteness through a very specific, even if quite obscure, debate about methodology. The methodological preference for complexity among those in the field of environmental migration research may signal a new ontology and, indeed, a new methodology. But, as I have tried to show in this chapter, this complexity turn is neither race-neutral nor new but conforms methodologically to the dominant mode of the contemporary racist expression of "not racism."

In the next two chapters, I examine how a new form of racial rule—racial futurism—is taking shape alongside and within the adaptation of capitalist political economy to climate change. Just as the nineteenth-century environmental determinism formed part of the ideological scaffolding for European imperialism and commodity capitalism, I examine how the discourse on climate change and migration appears to be doing the same thing but for *contemporary* global capitalism, whose distinctive historical form is no longer commodity production but financial speculation.

McLeman and Gemenne are very clear about the fact that the nineteenth-century environmental determinism was sustained by "dubious theoretical propositions, uneven methodological rigor, racist undertones, and *de facto* legitimation of colonialism."[70] What they miss, however, by neglecting to account for the history of race and racism, is that the contemporary discourse on climate change and migration also contains dubious theoretical propositions, uneven methodological rigor and racist language, all of which play a role in the adaptation of capitalist political economy to climate change. As we will see over the next two chapters, forms of knowledge that seeks to anticipate future "climate migration" in advance of its occurrence are at the heart of this endeavour.

## NOTES

1. Ta-Nehisi Coates, *Between the World and Me* (Melbourne: Text Publishing 2015), 97.

2. Reni Eddo-Lodge, *Why I Am No Longer Talking to White People about Race* (London: Bloomsbury, 2018), xi.

3. Alana Lentin, "Beyond Denial: 'Not Racism' as Racist Violence," *Continuum* 32, no. 4 (2018): 400–14; Alana Lentin. *Why Race Still Matters* (London: Polity, 2020).

4. Lentin, "Beyond Denial," 402.

5. David Theo Goldberg, *Are We All Postracial Yet?* (Cambridge: Polity Press, 2015).

6. I borrow the phrase "move to innocence" from Eve Tuck and K. Wayne Yang, "Decolonization Is Not a Metaphor," *Decolonization: Indigeneity, Education and Society* 1, no. 1 (2012).

7. Gurminder Bhambra, "Brexit, Trump, and 'Methodological Whiteness': On the Misrecognition of Race and Class," *The Journal of British Sociology* 68, n.S1 (November 2017): S214–S232.

8. Lentin, *Why Race Still Matters*, 56.

9. On the concept of state racism, see David Theo Goldberg, *The Racial State* (Malden: Blackwell, 2002). A full list of texts that address the historical links between racism and immigration is beyond what I can provide here. For a very brief list, see, for example, Lake and Reynolds, *Global Colour Line*; Panikos Panayi, *An Immigration History of Britain: A Multicultural History of Racism since 1800* (New York and London: Routledge, 2010); Mekkonen Tesfahuney, "Mobility, Racism and Geopolitics," *Political Geography* 17, no. 5 (1998): 499–515; and Barrington Walker, ed., *The History of Immigration and Racism in Canada: Essential Readings* (Toronto: Canadian Scholars Press, 2008).

See also Angus McLaren, *Our Own Master Race: Eugenics in Canada, 1885–1945* (Toronto: University of Toronto Press, 2015) on the historical linkages between immigration control and eugenics.

10. Lentin, *Why Race Still Matters*, 62.

11. While in recent years #BlackLivesMatter has done much to expose how race and racism are endemic features of everyday life in the West, race denialism continues to shape Western political discourses on race and racism.

12. Hesse, "Racialized Modernity"; Wynter, "Unsettling the Coloniality of Being/Power/Truth/Freedom"; Lisa Lowe, *The Intimacies of Four Continents* (Durham, NC, and London: Duke University Press, 2015).

13. Tuck and Yang, "Decolonization Is Not a Metaphor," 3.

14. Wiegman, "Whiteness Studies."

15. Meyer and Guss, *Neo-Environmental Determinism.*

16. Wiegman, "Whiteness Studies," 119.

17. Castree, *Nature*.

18. Richard Peet, "The Social Origins of Environmental Determinism," *Annals of the Association of American Geographers* 75, no. 3 (1985).

19. Meyer and Guss, *Neo-Environmental Determinism*. For more on environmental determinism, see Peet, "Social Origins"; Andrew Sluyter, "Neo-Environmental Determinism, Intellectual Damage Control, and Nature/Society Science," *Antipode* 35 (2003): 813–17; David Livingstone, *The Geographical Tradition: Episodes in the History of a Contested Enterprise* (Oxford: Blackwell, 1992); David Livingstone, "Environmental Determinism," in *The Sage Handbook of Geographical Knowledge*, ed. John Agnew and David Livingstone (Thousand Oaks, CA: Sage, 2011), 368–80;

and Sarah Radcliffe, Elizabeth Watson, Ian Simmons, Felipe Fernández-Armesto and Andrew Sluyter, "Environmentalist Thinking and/in Geography," *Progress in Human Geography* 34, no. 1 (2010): 98–116.

20. David Livingstone's *The Geographical Tradition* offers a good starting point.

21. See, for example, Colin Kelley et al. "Climate Change in the Fertile Crescent," as well as Jared Scott's popular American documentary film *The Age of Consequences* (Moon Atlas and PF Films, 2016), 80 mins.

22. U.K. Foresight, *Migration.*

23. For a useful account of the political discourse on security, climate change and migration, see White, *Climate Change and Migration.*

24. Richard Black and Michael Collyer, "Populations 'Trapped' at Times of Crisis," *Forced Migration Review* 45 (2014).

25. Bettini, "Climate Migration as an Adaption Strategy"; Chris Methmann and Angela Oels, "From 'Fearing' to 'Empowering' Climate Refugees: Governing Climate-Induced Migration in the Name of Resilience," *Security Dialogue* 46, no.1 (February 2015); Andrew Baldwin, "Resilience and Race, or Climate Change and the Uninsurable Migrant: Towards an Anthroporacial Reading of Race," *Resilience: International Policies, Practices, and Discourses* 5, no. 2 (2017); Felli, "Managing Climate Insecurity"; Romain Felli and Noel Castree, "Neoliberalising Adaptation to Climate Change: Foresight or Foreclosure?" *Environment and Planning A* 44, no. 1 (2012).

26. U.K. Foresight, *Migration and Global Environmental Change*, 11. Emphasis is mine.

27. Meyer and Guss, *Neo-Environmental Determinism*. Richard Peet's ideological critique appears some sixty years after the initial wave of critiques broke in the 1920s. See Peet, "Social Origins."

28. Meyer and Guss, *Neo-Environmental Determinism.*

29. Asian Development Bank, *Addressing Climate Change and Migration in Asia*, 1.

30. McLeman, *Climate and Human Migration.*

31. Ibid., 9

32. Piguet, "'Primitive Migration' to 'Climate Refugees,'" 149.

33. François Gemenne, "Migration Doesn't Have to Be a Failure to Adapt: An Escape from Environmental Determinism," in *Climate Adaptation Futures*, ed. Jean Palutikof, Sarah Boulter, Andrew Ash, Mark Smith, Martin Parry, Marie Waschka and Daniela Guitart (Oxford: Wiley-Blackwell, 2013), 235–41.

34. Ibid., 235–36.

35. Additional examples include François Gemenne et al., "Climate and Security"; Gemenne, "4 Degree Celsius+ World"; Lori M. Hunter, Jessie K. Luna and Rachel M. Norton, "Environmental Dimensions of Migration," *Annual Review of Sociology* 41, no. 1 (2015); Graeme Hugo, "Future Demographic Change and Its Interactions with Migration and Climate Change," *Global Environmental Change* 21 (2011).

36. Nash, *Negotiating Migration.*

37. The U.K. Climate Change Migration Coalition has been at the forefront of efforts to persuade media outlets worldwide to abandon alarmist claims when

reporting on the relationship between climate change and migration. https://climatemigration.org.uk/.

38. Lentin, *Why Race Still Matters*, 58.

39. Ladelle McWhorter, *Racism and Sexual Oppression in America: A Genealogy* (Bloomington: Indiana University Press, 2009).

40. Ibid., 36.

41. Lentin, *Why Race Still Matters*, 52–92.

42. Gordon Manley, "Some Recent Contributions to the Study of Climate Change," *Quarterly Journal of the Royal Meteorological Society* 70, no. 305 (July 1944): 197–98.

43. Piguet, "'Primitive Migration' to 'Climate Refugees,'" 158.

44. To be fair, in their editor's introduction to the *Routledge Handbook on Environmental Displacement and Migration*, McLeman and Gemenne offer an abridged critique of environmental determinism and make a point of arguing that "environmental determinism was discredited by other scholars for its dubious theoretical presuppositions, uneven methodological rigour, racist undertones, and de facto legitimation of colonialism." See McLeman and Gemenne, "Environmental Migration Research," 8.

45. Gemenne, "Human Face."

46. Dun and Gemenne, "Defining,"10.

47. Gemenne et al., "Climate and Security."

48. Campbell et al., *Age of Consequences*, 72. Emphasis in original.

49. Gemenne, "Human Face."

50. Kelley et al., "Fertile Crescent," 3245. My emphasis.

51. Gilroy, *Darker than Blue*.

52. Ellsworth Huntington, *Civilization and Climate*, third edition (London: Oxford University Press, 1924), 27.

53. Lentin, "Beyond Denial"; Lentin, *Why Race Still Matters*.

54. Goldberg, p*ostracial*. See also, Eduardo Bonilla-Silva, *Racism without Racists: Color-Blind Racism and the Persistence of Racial Inequality in America* (Lanham, MD: Rowman and Littlefield, 2018).

55. Goldberg, p*ostracial*, 109.

56. Lentin, *Why Race Still Matters*, 55.

57. Ann Laura Stoler, *Duress: Imperial Durabilities in Our Times* (Durham, NC, and London: Duke University Press, 2016),12. My emphasis.

58. Lentin, "Beyond Denial," 401.

59. Lentin, *Why Race Still Matters*, 5.

60. Ibid., 6.

61. Ibid., 66.

62. Wynter, "Unsettling."

63. Wiegman, "Whiteness Studies," 121.

64. Ibid., 119.

65. Ibid., 120.

66. Ibid., 120.

67. Ibid.

68. James Morrissey, "Rethinking the 'Debate on Environmental Refugees': From 'Maximalists and Minimalists' to 'Proponents and Critics,'" *Journal of Political Ecology* 19 (2012).

69. Hulme, "Reducing," 247.

70. McLeman and Gemenne, "Environmental Migration Research," 8–9. It is noteworthy that these authors refer to the "racist undertones" of early environmental determinism, when, in fact, the racism in this literature is really quite explicit. Lentin has identified how the muted reference to racism through terms such as "racial animosity," "racial attacks" and "racial connotations" amounts to a recognisable genre of "not racism" that she calls the "euphemization of racism." See Lentin, *Why Race Still Matters*, 52.

# Chapter 5

# Premediating Climate Change and Migration

*Postcards from the Future* is a collection of photomontage images exhibited at the Museum of London and the National Theatre in 2010–2011.[1] Created by U.K. artists Robert Graves and Didier Madoc-Jones, the sixteen images that comprise the collection are an aesthetic interpretation of London adapting to climate change. Migration is one of the exhibition's core themes. In one image, Buckingham Palace is engulfed by an informal settlement. In another, the double glazing in London's famous Gherkin is draped with drying laundry. In a third, Parliament Square is transformed into a rice patty, worked on by hunched over workers and a yoked water buffalo, while in a fourth image, a family of Jaipur monkeys overlook the flooded Thames from their refuge atop St. Paul's Cathedral. Race is nowhere stated in these images and yet they scream race: the institutions of Britishness, of whiteness—monarch, capital, parliament and church—all overrun by nameless climate migrants.

*Postcards* reflects a Euro-western fascination with climate change futures. In this chapter, I use the collection as another site from which to theorise racial futurism. *Postcards* has been interpreted in a variety ways. It was celebrated by Vivian Westwood, Norman Foster and Radiohead as an effective call to action. But it also came under heavy criticism for reinforcing migrant stereotypes. However, it is Martin Mahony's interpretation that best captures how the collection is structured by racial futurism.[2] As he puts it, the images offer "a simplistic racialised politics of place."[3] But he also says they leave "the overall impression of a stoic city—resilient, adaptable, like the city narrated as a 'survivor' of the great fire of 1666 and of the bombs of World War II."[4] Mahony's juxtaposition of race and urban adaptation is important. It calls attention to a long-standing narrative of London as a proud imperial capital capable of adapting to whatever hardships history throws its way. While at the same time it suggests that governing "climate refugees" will play an important part in London's future adaptation to climate change. But if we look at the images carefully, what we see is not a city run amok, but one in which the figure of climate migrant/refugee is an ordered presence in the city. London may appear to be inundated by "climate refugees," and yet somehow it remains strangely the same.

*Postcards* is a jarring exhibit because it ruptures our sense of time and space. The future it depicts is not somewhere "out there" on the horizon of time. *Postcards* makes the future present, weaving the future into the here and now of London. Nor is the other of climate change "over there" on the outer peripheries of the metropole—in Africa, Asia, Latin America or the islands of the sea. It, too, is given a real presence in the city. By collapsing the distance between present and future, near and far, self and other, Graves and Madoc-Jones provide a vivid example of the spatial logic of racial futurism, a logic in which the unfathomable, incalculable other of climate change is made real, but a real which is virtual as opposed to one that is actual. If the other is that which cannot be known, then *Postcards* translates the other of climate change from the realm of the unfathomable into the realm of knowledge. The other becomes knowable to us, even if our knowledge of it remains a matter of speculation.

When we turn to the international political discourse on climate change and migration, we find similar speculations on space, time and difference. It is these that interest me in this chapter. International institutions of various sorts all rely on anticipatory epistemologies, such as vulnerability mapping, scenarios and modelling, to render the other of climate change real. Like Graves and Madoc-Jones's photomontage, techniques such as these are what create the powerful illusion that the figure of the climate migrant/refugee has objective existence even though, as we saw in chapter 2, there is a good reason to the contrary. In this chapter, I examine how one such technique—scenario planning—works to reify the other of climate change through the virtual spatialisation of Blackness, how scenarios nurture whiteness by cultivating *potential* "climate migration" *as* Blackness. By seeking to anticipate where and how climate change will affect migration, anticipatory forms of knowledge, such as scenarios, are said to hold the key to governing the looming migration crisis of climate change. The appeal of this kind of knowledge lies in the way it promises to pre-empt what many say will be the most catastrophic effects of climate change: political violence and humanitarian crises catalysed by migration.[5] From the perspective of liberal humanism, it is easy to grasp the appeal of scenarios, especially if they promise to safeguard vulnerable groups or curtail mass migration into already overcrowded cities. It is important to remember, however, that anticipatory knowledge of the future, like all forms of knowledge, even if well intentioned, is never neutral. As Ben Anderson has argued, when the "future becomes cause and justification for some form of action in the here and now," the logic of anticipation becomes political.[6] The same is true of knowledge about future climate migration. Such knowledge, not least scenarios, is vested with the power to intervene in the present in ways that stand to reconfigure everyday life for millions, possibly even billions, of people. Scrutinising these techniques is, therefore, of utmost importance in order to grasp racial futurism.

My argument is that through its reliance on scenarios, the international discourse on migration and climate change functions as a particular kind of security apparatus, one which generates an orientation to climate change I refer to in this chapter as "white affect."[7] My contention is that the power of these scenarios, the power of white affect, lies in the way they repeat the "operative logic" of anti-Blackness as a constitutive dimension of humanism.[8] This is of fundamental importance for any critical account of the discourse for at least two reasons. First, examining white affect will allow us to consider how anti-Blackness organises the discourse through the cultivation of a specific kind of feeling and sense. Anti-Blackness, in this affective sense, is not the overt expression of ignorance or hatred we might typically associate with racism. For Sexton, it is a kind of "structure of feeling" that operates without drawing attention to itself as racism. Anti-Blackness can be mobilised this way because affect need not be spoken to have an effect. Rendering white affect visible, making it the object of our analysis, can, however, help us grasp the persistence of race thinking in our supposedly postracial times.[9] Indeed, as David Theo Goldberg once remarked, one of the great difficulties in challenging racism in the historical context of "not racism" is identifying racism's "terms of account" when these have been "rendered invisible." By examining white affect, we stand to appreciate how racist sensibilities are mobilised for political purposes without the vocabulary of race ever being mentioned.[10] This will prove to be especially important in chapter 6 when we consider how anti-Blackness structures a specific variation of international discourse on climate change and migration scripted in the seemingly benign language of complex adaptative systems, risk and resilience. These are terms not normally associated with racism, and, yet, as I will argue, they remain fundamental to the governing logics of racial futurism. The second reason tracing anti-Blackness through white affect is important for critically evaluating climate migration discourse is that doing so calls attention to the way anti-Blackness operates as a pre-emptive form of power. The scenarios I discuss below pre-empt the future by confining the climate change imaginary to a narrow range of possible migratory futures. The scenarios, then, guide forms of political intervention in the present that seek to annul those futures from materialising, futures that, if allowed to materialise, would destabilize racial nomos. The concept of white affect can help us understand how anti-Blackness is at stake in this form of power.

Even more powerfully, though, attending to white affect can help clarify how pre-emptive anti-Blackness constrains the climate change political imaginary. White affect does this by foreclosing *other* ways climate futures might be imagined. For example, white affect cultivates a liberal humanism in which governing climate change adaptation amounts to rescuing the world's poorest and most vulnerable from the worst of climate change. This might involve, for example, supplying the most vulnerable with climate risk insurance or institutionalising legal rights for the so-called climate migrants or climate refugees. And while these interventions might sound reasonable to some, they foreclose *other* ways climate change might be politicised, such as using climate change as a site for contesting and reversing global regimes of migration and mobility structured by race, challenging the dominance of finance and insurance capital in climate change governance, or, indeed, using climate change as a political space for redefining humanism.[11]

As an object of political intervention, white affect deadens the climate change political imaginary by excluding structural racism from the terrain of the thinkable. However, by naming white affect, we stand to better appreciate how the anti-Blackness of climate migration is waged not simply through discourse and representation but on and through affect. So, in this respect, while I am broadly sympathetic with those who would argue that climate change, in general, and migration, in particular, have become depoliticised, it is also *here* that I part company with such arguments to the extent they disavow race, or indeed anti-Blackness, as a conceptual frame of reference for critically evaluating the political significance of the discourse on climate change and migration.[12] My concern is not so much that climate-migration scenarios are forms of knowledge that depoliticise the future.[13] What concerns me is that such scenarios are inherently political devices that white humanism actively wields at the scale of the international in order to claim the future for itself, tools which make the future uniquely available to whiteness for the sole purpose of strategising *its* adaptation.

Some readers will understandably find this idea difficult to accept. My argument, after all, appears to close down the possibility of assisting those who stand to be displaced by climate change. However, if we follow Frank Wilderson for whom anti-Black violence is *the* structural antagonism that founds white humanism, my argument may start to make more sense. Wilderson arrives at his position through the history of transatlantic slavery.[14] For him, and others writing in the genre of Afropessimism, slavery is to be understood not merely as an episode of the past but as an ongoing relation of violence that rests on "a symbiosis between the political ontology of Humanity and the social death of Blacks."[15] For these writers, it is the enduring legacies of chattel slavery that are said to define the relation between humanism and those barred access to the category of the human. This is a relation in which those positioned outside humanism are imagined to be available *for* humanism, to be defined, possessed and deployed in whatever way humanism sees fit. To the extent the figure of the climate migrant/refugee is itself defined outside the category of the human, then it, too, inhabits that same structural position as the Black subject. If we take Wilderson seriously, and I argue we should, then the discourse on climate change and migration starts to look more and more like a relation of violence which shares much in common with his political ontology of humanism.[16] To be clear, I am *not* arguing that those defined as would-be climate migrants or climate refugees are in fact slaves. This is clearly not the case. Rather, my point is that the figure of the climate migrant/refugee occupies the same structural position as the figure of the Black in Wilderson's political ontology of humanism. The figure of the climate migrant/refugee is a figuration of living death and nothingness that serves as the foundational condition *for* humanism, in order for the human to adapt to climate change as human. Any attempt to trace how humanism adapts to climate change therefore requires, in my view, that we confront *this* political ontology. In this chapter, I endeavour to show how the climate-migration relation repeats the political ontology of slavery, serving as the ground on and through which white humanism's anti-Black violence is learning to adapt to climate change. White humanism remakes itself in the crucible of its undoing by conjuring the virtual presence of Blackness. Or as Stefano Harney and Fred Moten put it, "Maybe we could say [governance] is provoked by the *communicability of unmanageable racial and sexual difference*."[17] This is racial futurism.

## THE POLITICAL APPARATUS OF WHITE AFFECT

Michel Foucault once described an apparatus as "a thoroughly heterogeneous ensemble consisting of discourses, institutions, architectural forms, regulatory decisions, laws, administrative measures, scientific statements, philosophical, moral and philanthropic propositions." "Its major function," he tells us, is "responding to an urgent need. The apparatus thus has a dominant strategic function."[18] To conceptualise climate change as a looming migration crisis within the scope of Foucault's definition of an apparatus, therefore, means approaching this problematisation as an historically specific strategic response to the climate change crisis. Drawing Foucault's notion of the apparatus into dialogue with racial futurism, my argument in this chapter is that the strategic deployment of the discourse entails the "strong conduction" of white affect. By which I mean, following Derek Hook, "a streaming or encouragement of certain affective bonds which . . . retain powerfully racializing elements."[19] My claim here extends the argument developed over the preceding chapters, especially chapter 2, which is that the figure of the climate migrant/refugee serves as the basis for white renewal—the adaptation of whiteness. Specifically, I examine how the discourse consolidates a "white" sensibility when the figure is *sensed* as a potential disruption to the ordinary.

The concept of white affect encompasses the dispositions, attitudes and sensibilities generated by the production and circulation of anticipatory knowledge convened within the apparatus of climate change and migration. So, when Giovanni Bettini argues that the discourse on climate change and migration produces a surplus of meaning that sutures multiple constituencies, such as militarism, humanitarianism, neoliberalism, ethnonationalism and developmentalism, through a shared sense that the discourse "sounds right," he comes close to identifying white affect.[20] Each of these various constituents may know the "crisis" differently, just as each may respond differently. Yet all can be understood to share the same *sense* that climate change is a looming migration crisis. This "sounding right" of the discourse is akin to what I am calling white affect, a kind of feeling or embodiment resonant with whiteness as a regime of power. When climate change is said to be a looming migration crisis, this is often because it threatens to dislodge the governing apparatus of whiteness; it threatens to undo racial nomos. *Postcards* is only just one example where we can locate the conduction of white affect. *The Atlas of Environmental Migration* discussed in chapter 3 is another.[21] This chapter examines two additional texts for the way each generates white affect through the problematisation of climate change and migration. Each is drawn from a distinctive political domain—security and development, respectively—and each uses scenarios to construct the relation between climate change and migration through the scale of the international. The first is *The Age of Consequences: The Foreign Policy and National Security Implications of Global Climate Change* published by the Center for Strategic and International Studies and the Center for a New American Security in 2007.[22] The second is *Groundswell*, published over a decade later by the World Bank, discussed in chapter 3.[23]

The concept of affect has gained considerable popularity across the social sciences and humanities in recent years. Although a contested concept, my use of it comes from Brian Massumi for whom affect is a synonym for intensity but also the term he uses to describe "*the virtual as a point of view.*"[24] By establishing white affect as my main object of analysis, my argument in this chapter departs from the dominant way of conceiving race as an effect of signification. I explain white affect in more depth later. For now, though, my use of white affect owes more to Arun Saldanha's observation that "white racism will need to be conceived as a system involving not just exclusion, but more complex shades of differentiation and interaction *prior to* any distinction between self and other, West and East."[25] My suggestion here is that white affect is ontologically prior to racism, prior to humanism's violent repudiation of Blackness. This is not to displace our understanding that racism is a system of meaning. It is simply to emphasise white racism's ontogenetic paranoia, which is only subsequently translated into a signifying power. White affect should, therefore, not be confused with something called white identity, which is the embodiment of a racist system of meaning.[26] Nor is white affect akin to whiteness, which I take to be a system of representation and material power. Instead, white affect should be understood as an intensity, "the virtual as a point of view," a relation between motion and rest, between process and its cessation.[27] We can think of white affect as fully embedded in the material organisation of contemporary life in the West, but as its immaterial counterpart, much like the kinetic potential lodged in a weapon or a stone sitting on a cliff edge.[28] Or we can think of white affect as the immaterial force which draws racial capitalism forward much in the way the vacuum on the leeward side of a sail propels the ship through the aerodynamic effect of pull. And here, I should add that when I refer to white affect as functioning prior to white representation, my intention is not to naturalise white affect, whiteness or white identity; affect is itself an artefact of power.[29] Its "strong conduction" is both strategic *and* political. My use of white affect is merely meant to convene our understanding of white racism onto the plain of the ontogenetic. If racism is a structural form of power, my point here is to suggest that it operates structurally not only through institutions (e.g., policing, settler colonialism, migrant detention) or meaning (e.g., languages, images, texts) but also through feeling. White racism is embodied and felt as much as it is material. Hopefully, my reading of white affect can clarify how racism intensifies as a structure of feeling on the ontogenetic plain of climate change.

## SCENARIOS OR BLACKNESS AS DISPERSED POTENTIAL

Scenario planning is an anticipatory methodology in which a range of possible futures are identified for the purposes of planning. As a mechanism for rendering the future knowable, scenario planning has become a standard feature of contemporary security culture since 9/11.[30] First developed by the futurist Herman Kahn in the 1950s, and later refined by petroleum interests in the context of the 1973 oil crisis, scenario planning is widely considered an effective means for anticipating and, thus, preparing for risk events that fall outside the scope of prediction.[31] Unlike predictive reasoning, which seeks to identify the most likely outcome of a set of conditions, thereby eliminating or at least minimising uncertainty, scenario planning is primarily concerned with improved decision-making while fully *embracing* uncertainty.[32] As a form of risk management, scenario planning differs fundamentally from insurance, which calculates the probability of a risk event on the basis of historical frequency.[33] Scenario planning, by contrast, attempts to know the future by approaching the present as a time-space which contains an infinite number of possible futures. Scenario planners select from the present a narrow range of future scenarios which are then used to anticipate and plan for all manner of future uncertainties. Scenarios thus allow for informed actions to be taken in the present, actions that aim to ensure effective governance should one or any combination of eventualities come to pass. However, because scenarios delimit the future in accordance with a set of prior assumptions, they are never objective readings of the future but will always issue from a particular subject position whose principal concern is its own self-maintenance in relation to future risk. This does not make scenario planning intrinsically wrong. It simply means that however useful scenarios may be, their usefulness is prefigured by the interests they are designed to serve. Scenario planning is, like Foucault's apparatus, a strategic devise.

Scenario planning is a technique now commonly used to govern climate change adaptation.[34] It is central to the anticipatory logic of the IPCC and used extensively by all manner of international institutions, including, as we will see momentarily, the World Bank. As Melinda Cooper put it some time ago, scenario planning "is perhaps the most ubiquitous and most consequential of epistemologies in contemporary politics."[35] Perhaps not surprisingly, it is also one now routinely used to reify the figure of the climate migrant/refugee. One early example can be found in the 2003 U.S. Pentagon report titled *An Abrupt Climate Change Scenario and Its Implications for United States National Security*, often said to be one of the founding texts in "climate security" discourse.[36] As the title suggests, *An Abrupt Climate Change Scenario* presents just a single, albeit rather extreme, future: the abrupt slowing of the oceanic thermohaline conveyor.[37] The authors of the report are very clear, however, that the security implications they identify are not meant to be objective. Rather, the stated purpose of the report is to "imagine the unthinkable" as a necessary step in preparing the United States for abrupt climate change. This injunction to think the unthinkable would later be mainstreamed in U.S. security practice after the U.S. 9/11 Commission found that the events of 9/11 owed partly to the failure of U.S. security institutions to anticipate them.[38] We should, therefore, read *An Abrupt Climate Change Scenario* not as a discrete, one-off text about climate change, but as a reflection of contemporary U.S. security culture.

*An Abrupt Climate Change Scenario* presents a far-fetched scenario in which global resource depletion, escalating geopolitical tensions and violence combine to threaten U.S. imperial hegemony. Somewhat surprisingly, refugees figure minimally within the main body of the report. They do, however, appear in its conclusion:

> Large population movements in this scenario are inevitable. Learning how to manage those populations, border tensions that arise and the resulting refugees will be critical. New forms of security agreements dealing specifically with energy, food and water will also be needed. In short, while the US itself will be relatively better off and with more adaptive capacity, it will find itself in a world where Europe will be struggling internally, large numbers of refugees washing up on its shores and Asia in serious crisis over food and water. Disruption and conflict will be endemic features of life.[39]

Three points can be drawn from this passage that can usefully inform our discussion about white affect. First, imagining the unthinkable is a form of strategic practice. It is central to the way powerful institutions like the U.S. military strategise *their* endurance. The authors of the report readily admit that even while plausible, the scenario it presents is rather unlikely. So why engage in such speculation? Their answer is that "rather than predicting how climate change will happen," the purpose of imagining the unthinkable "is to dramatize the impact climate change *could* have on society if we are unprepared for it" and "to further the strategic conversation."[40] Second, imagining the future of the figure of the climate migrant/refugee produces a rather curious geography. Like *Postcards*, it does not depict the figure on the horizon of linear time. According to the logic of the scenario, it says that the climate migrant/refugee is *already present* in the present, albeit in a dormant, unactualised state. In this sense, the report conceptualises the figure as a form of dispersed immanent potential as opposed to an object that is already fully formed. This is what I mean by the virtual spatialisation *of Blackness*. The figure is real, but its realness is a function of its plausibility and potentiality, as opposed to its objective measurability. The figure is, in this sense, real but not actual. Here, we might think of the figure as Blackness spatialised in the affective idiom of the yet-to-come. And finally, the scenario is significant because it invokes the figure's dormant potential—its yet-to-come-ness—to bring about the strategic renewal of the U.S. military. The scenario is, thus, not about prediction but readiness. Not about preparing to fight a known enemy but reorienting military planners to a future in which the figure of the climate migrant/refugee is imagined as a source of geopolitical instability. In the context of a discussion about the apparatus of U.S. security in relation to climate change, we can think of the figure of the climate migrant/refugee as the *virtual* presence of Black social death, an improbable future conjured purposefully to bring about the strategic renewal of the U.S. military. This is what Wilderson means when he says that Blackness can be harnessed as "a resource for Human renewal."[41]

In the following two sections, we can see how *The Age of Consequences* and *Groundswell* use scenarios to locate the dispersed virtual potential of Blackness.

## *The Age of Consequences*

*The Age of Consequences* was published in November 2007, not long after Al Gore captivated audiences worldwide with the release of *An Inconvenient Truth* and around the time Gore and the IPCC would receive the Nobel Peace Prize. The UN Security Council would consider climate change for the first time earlier that year in April 2007, while the influential *Stern Review on the Economics of Climate Change* had come out just a year earlier.[42] It was during this period that the strong resonance of climate change science, policy, security and cinema would secure for climate change a lasting place in the West's popular imaginary. This moment was partly constituted by *The Age of Consequences*.

A formative text in the genre of "climate security," it is organised primarily around three climate change scenarios, which are labelled in the report "expected," "severe" and "catastrophic." The scenarios are based on projections of global surface temperature increases and sea-level rise and are used, in the text, by U.S. security experts to speculate on the national security implications of climate change over a 30- to 100-year period. All three scenarios identify climate-induced migration as a catalyst for political violence or at least geopolitical instability. They anticipate, for example, how water shortages in Mexico "will drive immigration into the United States,"[43] how those living on Bangladesh's southern coast will be forced to move to Dhaka or Chittagong or, worse, create regional instability by relocating to India, and how Muslims migrating from Africa to Europe could "increase the likelihood of radicalization among members of Europe's growing (and often poorly assimilated) Islamic communities."[44]

That the scenarios in *The Age of Consequences* are structured by a racist geopolitical imaginary is self-evident. That this imaginary is anti-Black may not be. Migration is imagined across all three scenarios as a pending chaos, culminating in the breakdown of an order of governance in which U.S. imperial power is pushed to the existential brink. But from Jared Sexton, we can interpret this pending chaos *as* Blackness. Blackness is not reducible to Black experience or Black history or even phenotype. For Sexton, it is an ontological presence not just outside every social bond but that "which relates to the undoing or unraveling of every social bond."[45] Or, as Wilderson puts it, Blackness is "a programme of complete disorder."[46] If the human is the subject that knows, speaks and reasons, Blackness exceeds these actions. It is the unspeakable and unknowable presence that grounds humanism.[47] It is true that as a means for political reflection Blackness originates conceptually from the history of American slavery. Its applicability for critical thought, however, extends well beyond its historical roots. Thus, it should hardly surprise us that the U.S. military would invoke anti-Blackness to structure a popular foreign policy text. Yet there remains an important distinction to be drawn between race and anti-Blackness. The former is a classificatory system that establishes the hierarchy to which the other is forced to conform. On this basis, the figure of the climate migrant/refugee is a racial category that reflects race. Blackness, by contrast, at least on this more ontological reading, is precisely that which precedes the racist impulse to order. It is the frightful excess, the provocative heterogeneity, to which racist categorisation is the answer. It would seem, then, that one of *The Age of Consequences*'s main strategic functions is to foment a sense of ontological terror, Blackness, which is then categorised as the figure of the climate migrant/refugee.

But the text also complicates my argument, introducing what appears to be a completely different other of climate change: the figure of the terrorist. In the catastrophic scenario, for example, climate change is said to have the capacity to unleash mass terrorism, which, in turn, is imagined to pose the *truest* threat to America. In a reaffirmation of Bush-era counterterrorism rhetoric, *The Age of Consequences* tells us that "malevolent" terrorists are said to want to "destroy our entire civilization and way of life." How, then, to make sense of this complication?

Andrew Telford's conception of "climate terrorism assemblages" is useful here. A "climate terrorism assemblage" is "a complex, emergent 'whole' formed from a heterogeneous range of interacting geopolitical components," including climate change impacts, migration, the discourse on "climate security," food insecurity, affective resonances and more.[48] What makes these assemblages important is their heterogeneous *emergent* potential not just their individual components. We could say the same of *The Age of Consequences*. It is significant not because it constructs the terrorist or the migrant or even environmental change as discrete objects we should fear but because *all* of these components intermix to produce the assemblage as *the* other of climate change.[49] What makes climate change so dramatic from the perspective of security is that it becomes indistinguishable from *all* of these imagined vectors of violence. All are dormant in climate change in equal measure. The terrorist, the migrant, the lethal virus, the rapist, the flood and so forth, all blur together as figurations of climate change in its most extreme, violent form. Together, they terrorise the human.

Christina Sharpe's meditations on Blackness connect up the themes of terror, migration and Black life in a way that allows us to recast "complex terrorism assemblages" as anti-Black technologies. It is worth quoting Sharpe at length:

> Living in the wake [of slavery] means living in and with terror in that much of what passes for public discourse *about* terror we, Black people, become the *carriers* of terror, terror's embodiment, and not the primary objects of terror's multiple enactments; the ground of terror's possibility globally. This is everywhere clear as we think about those Black people in the United States who can "weaponize sidewalks" (Trayvon Martin) and shoot themselves while handcuffed (Victor White III, Chavis Carter, Jesus Huerta, and more), those Black people transmigrating the African continent toward the Mediterranean and then to Europe who are imagined as insects, swarms, vectors of disease; familiar narratives of danger and disaster that attach to our always already weaponized Black bodies (the weapon is Blackness).[50]

Although there is nothing about climate change per se in Sharpe's account of body and terror,[51] her point is that even as Black people have become the targets of actual state terror (e.g., police brutality, death in custody, stop-and-search), Black *being* (as opposed to Black experience) is precisely that which underpins global discourses about terror and, by extension, *climate* terror. Thus, from Sharpe, we can say that the political ontology of slavery—the very relation of violence that weaponises Blackness—underwrites climate terror assemblages. From this critical perspective, the manifold ontological terror of climate change can be understood as a distinctive geohistorical manifestation of Blackness—planetary potentiality as the undoing of both the Earth and racial nomos. If, as Chakrabarty argues, climate change overturns the ontological stability of the human,[52] such a view is only possible if it assumes in advance some form of absolute difference which serves as humanism's foundational exclusion. It is precisely this foundational difference of Blackness that, I would argue, is immanent to humanism, internal to humanism as its necessary exclusion. This immanent difference is what allows me to draw an equivalence between Blackness and the other of climate change. Thus, I suggest we can reinterpret the scenarios of mass terrorism and migration in *The Age of Consequences* as the repetition of the ontological terror of Blackness. Not only do the scenarios construct future climate change as the dispersed potentiality of Blackness in varying degrees of severity, but they also redefine the human in relation to potentiality itself. From this perspective, the human is not simply the subject who knows, speaks and reasons, but the subject whose knowing, speaking and reasoning is directed towards taking possession of the potentiality of Blackness, of domesticating Blackness by ordering it through the classificatory system of race. My contention is that white humanism (a structure of power, *not* an identity) is striving to adapt to climate change by taking possession of and governing this potentiality.

The figure of the climate migrant/refugee is a racial category invented to bring order to this emergent potential. Inasmuch as scenarios are used not to predict the future but to better respond to all future eventualities, they are the key to this procedure. They allow the human to domesticate futurity through "precaution, prevention and pre-emption."[53] Chapter 6 considers how this same dispersed potentiality is captured and monetised through the financialisation of risk.

## *Groundswell*

*Groundswell* differs radically from *The Age of Consequences*. Whereas the latter militarises climate change, the former is embedded within the logics of international development. Its scenarios are more technical, more scientific, than *The Age of Consequences*, which is not surprising, given that *Groundswell*'s purpose is to bring about better development and adaptation outcomes. It is also decidedly focussed on what it calls "internal climate migrants," those whose migrations are said not to traverse territorial borders. This focus on "internal climate migrants" is significant for at least two reasons. First, it redefines the "problem" neatly within the jurisdiction of the nation state, giving it the appearance of being manageable within the confines of the state apparatus. And second, it sidesteps the overtly racist geopolitical connotations associated with transboundary migration. Chaos, disorder and excess give way in *Groundswell* to governability and containment, giving rise to a more muted intensity than *The Age of Consequences*. But even while the scenarios in *Groundswell* may feel more benign, less racist, they are no less implicated in the strategic conduction of white affect.

*Groundswell* provides policymakers with a logic for governing so-called internal climate migration. Scenarios are central to this logic. Using what it refers to as "anticipatory diagnostics," it combines distinct data sets—Shared Socioeconomic Pathways (i.e., moderate versus unequal development) with Representative Concentration Pathways (of greenhouse gas emissions and temperature increases)—to produce a series of three "internal climate migration" scenarios—*pessimistic*, *more inclusive development* and *more climate-friendly*. These scenarios are then fed into a gravity model which measures the relative attractiveness of two or more locations, along with climate change impact data, to gauge population changes at the scale of the nation state. The model is run multiple times for each scenario, allowing the modellers to infer the number of "internal climate migrants" at different scales by comparing the different outputs. *Groundswell* estimates the total number of internal climate migrants in aggregate and across each of the regions: 143 million "internal climate migrants" by 2050. The results of the modelling and analysis are presented in aggregate for three subregions—sub-Saharan Africa,

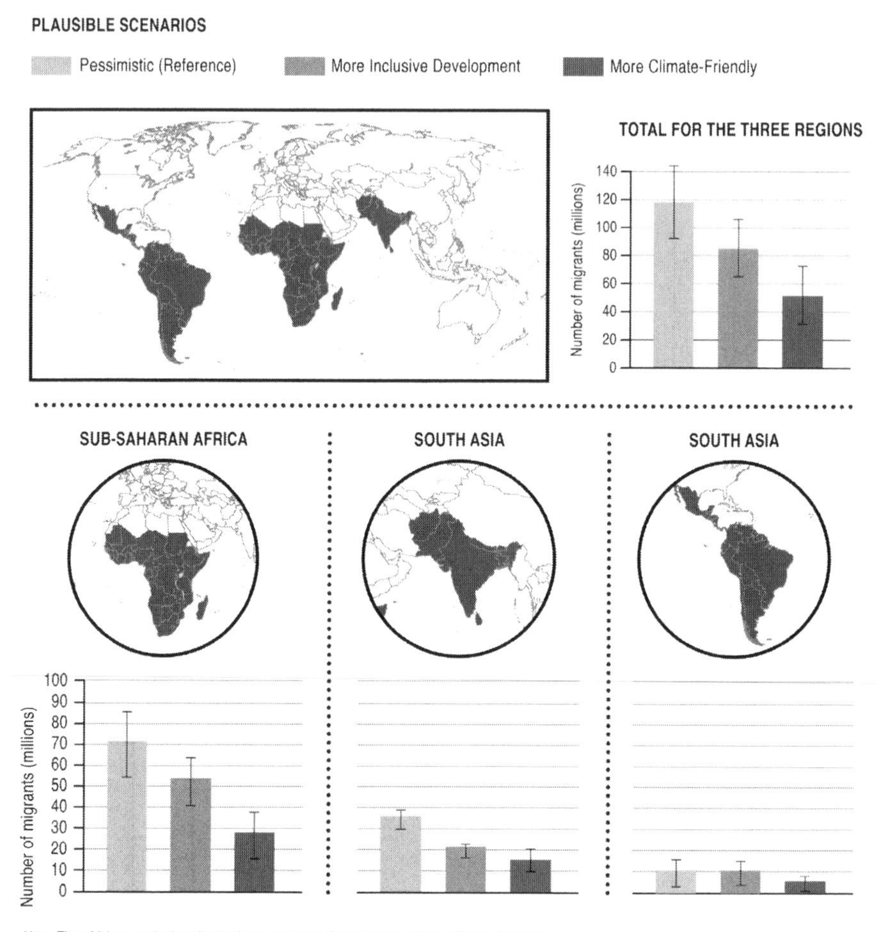


*Note: The whiskers on the bars in the charts represent the 95th percentile confidence intervals.*

**Figure 5.1 Infographic depicting "internal climate migration" scenarios for Sub-Saharan Africa, South Asia, and Latin America. Source: Image adapted from Rigaud et al., *Groundswell: Preparing for Internal Climate Migration.*[54]**

South Asia and Latin America (Figure 5.1)—and for three specific countries—Mexico, Bangladesh and Ethiopia. It also produces a series of maps of "climate in- and out-migration" hotspots, and net in- and out-migration for rural, coastal and urban areas in each subregion.

Evaluating the reliability of the model lies outside the scope of my expertise. It bears mentioning, however, that as far as climate migration modelling goes, the gravity model used in *Groundswell* represents a significant advance from first-generation modelling associated with Myers. As these models become increasingly more sophisticated, it is imperative that critical scholars find ways to scrutinise them, especially if our task is to understand how they circulate white affect and contribute to the adaptations of the political. The fact that *Groundswell* follows a rigorous method does not make it any less political than Myers's quantitative predictions or xenophobic texts like *The Age of Consequences*. Its scenarios may appear scientific, but we should remember that they issue from the World Bank, which continues to be one of the pillars of the global financial system. Understanding their strategic function cannot be separated from this important fact. Nor can it be separated from the fact that within post-9/11 security culture, development and security practices are not separate domains of state activity but increasingly understood to comprise two sides of the same apparatus.[55]

However, it is *Groundswell*'s performativity that offers the most productive means for grasping the power of white affect. By performative I simply mean that the text produces the effect it names.[56] We can think of performativity as a form of power that actively manufactures its object, reifying what is otherwise illusory or incalculable. One of the underlying assumptions of *The Other of Climate Change* is that there is no such thing as a climate migrant/refugee. The figure is nothing, an incarnation of nothingness, of Blackness. Its presence is sensed but the figure has no concrete existence.[57] This is the epistemic disjuncture that underpins the discourse on climate change and migration discussed in chapter 1. What the scenarios in *Groundswell* do, however, is draw the "internal climate migrant" into being by correlating proximate sets of data in a single model. The existence of the internal climate migrant is then inferred from the data.[58] Whatever realness it may be said to have is virtual not actual. Attending to this performative aspect of *Groundswell* is important because it allows us to conceptualise the "internal climate migrant" not as an unmediated, objective phenomenon but as the effect of a calculative apparatus put in train by a set of powerful international financial institutions. These calculative practices are what draw the incalculable figure of the climate migrant/refugee into virtual existence as a single numeric projection—143 million internal climate migrants by 2050. It is on this basis that we should approach *Groundswell* as a profoundly political text. It abstracts the actual everyday lives of those enumerated by the category "internal climate migrant" into a global calculus that cannot be separated from the strategic priorities of the World Bank and the wider global development-security-migration nexus it serves.

Nor can *Groundswell*'s performativity be decoupled from the "white gaze of development."[59] Much has been written about the way development discourse and practice are racially constituted.[60] It is not my intention to rehearse that argumentation here except to say that *Groundswell* follows a similar pattern of racialised knowledge formation whereby whiteness functions as a structure of power associated with authority, knowledge and expertise in development contexts. *Groundswell* is interesting here. It is certainly structured by the white gaze of development. Yet it differs fundamentally from conventional development narratives on account of its temporality. Conventional development is narrated in terms of progressive, linear time according to which those said to be in need of development are understood to progress from being "underdeveloped" or "developing" to "developed." This is a variation on Goldberg's conception of racial historicism. But in an important shift, and this is typical of climate change–development narratives more generally, *Groundswell* recalibrates the temporality of development's white gaze by constructing its other, in our case the internal climate migrant, in the language of risk. No longer is the other marked by some fundamental lack (e.g., of rights, capital, education or health) which development promises to overcome. From the vantage of *Groundswell*'s white gaze—the virtual as a distinctly *racialised* point of view—the other is imagined as a future risk event that *could* emerge at some point in the future, if actions are not taken in the present to pre-empt its emergence. In this way, *Groundswell* reorients development's white gaze in relation to a slowly unfolding global crisis in which its other is said to be *in*-formation. The double meaning of *in-formation* is deliberate.[61] It refers to the idea that the other is in the process of being formed or getting into formation. But it also refers to the idea that the other is nothing except information, an *inference* whose conditions of possibility are an arrangement of correlated datasets. Not yet fully actualised, but no less present even if only present in the data.

*Groundswell*'s "anticipatory diagnostics" has important implications for how we might reconceptualise the white gaze of development. As with *The Age of Consequences*, *Groundswell*'s racial other is not so much a distinctive body, or agent, as potentiality itself. The text's white humanism seems to be more concerned with governing potentiality, that is, governing the process of the other's formation, how the other comes into the world, than with addressing the other's perceived lack. Indeed, one of *Groundswell*'s key messages is that "internal climate migration may be a reality but it doesn't have to be a crisis."[62] Anticipatory diagnostics is, in this sense, only partly about anticipating where and how the crisis will unfold. It is also about governing the crisis by ensuring it does not actually materialise. In one illustrative passage, it notes:

> Where the limits of local adaptation are anticipated, well-planned migration to more viable areas can be a successful strategy. A strong enabling environment for migration needs to be in place supported by direct incentives, such as skills training and job creation programs, for people to move to areas of low risk and greater opportunity. Strategies supporting internal migration need to safeguard not only the resilience of those moving, but also of those in sending and receiving communities.[63]

To be clear, *Groundswell* is not a prescriptive text. It is simply meant to demonstrate how the gravity model can help policymakers anticipate and, thus, govern the "problem" by pre-empting its emergence. As with the earlier passage, this logic partly involves directing investments so as to annul the event of "internal climate migration" before it materialises. Take, for example, the case of Bangladesh, where the state is already quite advanced in building "climate migration" into adaptation planning. Under the pessimistic scenario, the model anticipates high out-migration from the agriculture-dominant northeast, high in-migration along the Ganges River in Western Bangladesh, and, somewhat counterintuitively, high-out migration from Dhaka and Chittagong. *Groundswell* does not instruct the Bangladeshi state how it should respond to the problem. It simply suggests that armed with this kind of anticipatory knowledge, development and planning authorities stand to make improved investments in infrastructure and resilience in advance of these patterns. A close reading of Bangladesh's *National Strategy on the Management of Disaster and Climate Induced Internal Displacement* suggests, for example, that one such investment might include "promoting and encouraging decentralization" of large urban centres like Dhaka, redirecting would-be migrants away from Dhaka to other areas of Bangladesh where labour is needed.[64] The scenarios in *Groundswell* become potentially useful for enabling this kind of pre-emptive governing. Although much can be said about "climate migration" policy in Bangladesh, the important point here is that policies such as urban decentralisation are conceptualised in a way that "drains" the actual event of "internal climate migration" of its "eventfulness,"[65] an example of "what could have been, but did not take place."[66] Accordingly, whatever migration that does result from the policy of decentralisation would have been the result of the policy not climate change. This does not make the policy inherently wrong. I raise it here simply to illustrate that it is informed by the logic of pre-emption.

There is nothing remotely novel about any of this. The logic of pre-emption is an endemic feature of the contemporary neoliberal security state. *Groundswell* simply outlines a modelling technique that could prove useful for those tasked with ensuring that climate change does not undermine development in the long run. As for the white gaze of development, what is important here is not whether anticipatory diagnostics are accurate or whether pre-emption generates its desired outcome. This is not a binary ethics of good or bad. What matters is that from *Groundswell* we can see how white humanism is consolidated in relation to the *potential* crisis of "internal climate migration." It is the relation of potential that produces white affect that is so central to racial futurism.

## PREMEDIATION

Premediation is a concept that can help us further grasp the way that the discourse on climate change and migration functions as an apparatus of white affect. Premediation is a concept used in media and communication studies to describe a form of American media logic which arose in the immediate aftermath of 9/11 and which intensified in the run- up to the U.S.-led invasion of Iraq in March 2003.[67] Coined by Richard Grusin, premediation is a kind of media logic that actively proliferates the number of possible future scenarios that may arise out of a given future event, such as war, an act of terrorism or climate change.[68] The purpose of premediating such futures—mediating them before they occur—is to inure publics to the shock of the event and its multiple aftermaths in advance of their occurrence. In this way, premediation functions as a kind of "affective prophylactic."[69] It is a technique that can be used to prepare publics to expect that the future will take any number of possible forms. We can think of *Postcards*, *The Age of Consequences* and *Groundswell* as examples of texts that all make use of the same logic of premediation. The first premediates climate change as a problem of urban adaptation and race. The second premediates climate change as a set of possible escalating crises, while the last premediates climate change as a series of migratory futures which, if properly managed, will not result in the kinds of crises foretold in the second.

Two specific elements of premediation are pertinent for our discussion. The first concerns how "premediation entails the generation of possible future scenarios or possibilities which may come true or which may not, but which work in any event to guide action (or shape public sentiment) in the present."[70] According to Grusin, what matters in premediative logic is not whether the scenarios it conjures are correct. Premediation, like scenario planning, is not a form of predictive reasoning. Rather, what matters is that by proliferating possible futures *before* they arise, premediation allows for the future to become actionable in the present,[71] for an action to be taken in the present based on a range of premediated futures. In this sense, premediation is not concerned with preventing premediated futures from occurring but with ensuring "that whatever form the future takes it will emerge only within the possible futures enabled by premediated networks of technical, social and cultural actors."[72] In this sense, premediation, like pre-emption, is a mechanism for domesticating the future, for ensuring that the future unfolds in a manner consistent with its premediation. The second element pertinent for our discussion is that premediation is centrally concerned with managing collective affect in such a way as to ensure that the future will not be experienced as a disruptive force but as a fully expected, even logical, outcome of the present. As Grusin puts it, "Premediation names both of these senses—the production of specific future scenarios *and* the creation and maintenance of an affective orientation towards the future, a sense of continuity or the feeling of assurance" that the future will unfold in the way we have been told it will.[73]

In the previous section, we saw how both *The Age of Consequences* and *Groundswell* premediate the future through the use of scenarios, albeit in starkly different ways. One is concerned with national security and the other with development. Both, however, premediate the future by naming the migration effects of climate change in advance of their occurring and, in doing so, at least partially ensure that, should such futures materialise, they will not register as unanticipated political emergencies but as fully anticipated outcomes of climate change. Each mediates the future before it occurs and, in this sense, each works as a kind of affective prophylactic by orienting its readership to the future through an anticipatory affect. Premediation can help us grasp how these texts contribute to the strong conduction of white affect.

Let me elaborate on what I mean by white affect. By affect I refer to an intensity that arises from the dynamic interplay between process and cessation,[74] and by white affect, I refer to an intensity that prefigures whiteness as a structure of power. We can think of white affect as a kind of intensification or closing down of the myriad futures available to bodies in a way that confines them to a set of constraints expressed as "whiteness." This requires further explanation, so let us start with the relation between process and cessation. Borrowing from the philosopher Brian Massumi, process can be thought of as the immanent and continuous unfolding of the totality of the world.[75] This is a view of the world in which process is given ontological primacy over position: before position (e.g., identity, subject, agent) we have process. A body in process is fully indeterminate. It has no determinate position but is instead best construed as pure potential. In this sense, a processual body has the capacity to become anything. However, a processual body stops being a processual body the moment it takes up a position, the moment it is stabilised, the very moment it encounters its own cessation. Thus, we might say that positionality is back-formed from cessation, back-formed against some termination point, for example, a crisis or an emergency. In this sense, when a body is back-formed against cessation, it undergoes a qualitative shift from being pure potential to being a determinate body, a positioned body, now defined, fixed or constrained within a limited set of possibilities.[76] This is not to say that the body stops moving (space and time continue to unfold); it is simply to say that the body's movement is no longer defined by its infinite variability. Rather, the body's movement is now defined by an affective relation between what the body is (e.g., Black, white, gay, straight, man, woman, and so on) and what it might become, between its actual (even if socially constructed) form and its virtual form. Back-formed against cessation, the body shifts from being a body in process to an intensive body. It moves from being a body marked by its potentiality, or infinite variability, to a determinate body unfolding within a framework of possibilities.

What I want to suggest is that white affect is a kind of intensity that arises when "white" positionality is back-formed against its own cessation, the fantasy of its own death, for example, or an ungovernable presence. To paraphrase Lake and Reynolds, white affect takes shape in the "apprehension of imminent loss."[77] White affect is, thus, not a processual subjectivity or one defined by infinite variability (even though whiteness remains highly mutable). White affect is an intensive or evental condition of the body, an affective intensity distilled from the ontogenetic interplay of back-formation when the infinitely variable body confronts the image of its own death. We can think of white supremacy as a specific intensity of white affect back-formed against imagined futures of "white replacement" and "race genocide." Similarly, we can think of white humanism as back-formed in the moment of encounter with its colonial other, the ambivalent feeling of emasculation and the fantasy of unlimited power that arise in the moment of imperial discovery.[78] In either case, whiteness is not foremost an effect of signification or a system of power that originates racial meaning. It is, first and foremost, an affective condition or intensity *prior to* the representational practices that make up the system of meaning we know as racism. Whiteness is, of course, a historically constituted structure of power, and to be sure the ontogenetic dynamics I describe here *are* historically situated.[79] Indeed, this dynamic is entirely consistent with the observation that "white identity" is a racial identity which imagines itself as the invisible backdrop against which racial difference is made to appear as such, or what Derek Hook describes as racism's "silent denominator."[80] My contention is simply that whiteness is preceded by an affective intensity. Whiteness may, of course, come to "consolidate a series of more explicit values,"[81] for example, a desire for control, innocence, salvation, purity, transcendence, particularity or the desire to save, all of which animate the discourse on climate change and migration. I would suggest, however, that these are signifying relations that derive from a prior affective intensity. Or as Hook suggests, whiteness is an "affective formation, a relational interplay of attractions and aversions, as a mode of subjectification that appears to exceed explicitly discursive forms."[82]

The discourse on climate change and migration is a premediative security apparatus whose political significance lies in the way it orients us to the future through a white affective intensity. In this sense, it would be incorrect to say that the discourse is simply xenophobic or that it mobilises racism to the task of combatting climate change. Racism is a discriminatory system that assigns value to people on the basis of biological and cultural markers of difference, and as far as I can tell the discourse has not yet been mobilised to promote actual instances of racist discrimination. So rather than discriminatory racism, I argue that the discourse produces a white racial orientation to climate change, one in which the future comes to be imagined on the verge of migratory disorder. By premediating the future as one in which the figure of the climate migrant/refugee, a figure synonymous with the structural position of Blackness, is said to be dormant in the present, climate migration scenarios like those contained in *Postcards*, *The Age of Consequences* and *Groundswell* contribute to a sense in which it has become completely normal to imagine the present as marked by potential climate-induced migration. When someone like the British prime minister, Boris Johnson, says that a failure to govern climate change could result in mass migration, we think nothing of it. Such an affective intensity is one that strives to normalise the assumption that nowadays all migration, either in the past, present or in the future, carries even just the slightest trace of climate change. These texts do this by placing the human, or humanity, in relation to the "what-could-be." They intensify this normal as a relation of possible transformation, such that the population, always assumed as a coherent whole, whether coded as "civilization," humanity, nation, community and so forth, and always assumed to possess a coherent set of values, is tensed on the verge of transformation. For example, if the phenomenon of climate change–induced migration is premediated through the proliferation of stories, images or scenarios of disordered or chaotic futures in which, for example, Jaipur monkeys will occupy St. Paul's Cathedral, or in which Buckingham Palace is engulfed by shanties, or in which the combined effects of climate change, Islamic fundamentalism and increasing northward migration from Nigeria to Tunisia threaten Europe's southern border, such images orient the viewer to a future in which whiteness cannot be easily guaranteed. In other words, such premediations anticipate the demise or disfiguration of white humanism. Or, as in the case of *Groundswell*, they premediate the future in a way that allows us to imagine how "internal climate migration" could but does not necessarily need to explode into a full-blown crisis. They tell us that such futures can and should be contained using the logic of pre-emption. Here, we might even say that *Groundswell* acts as a kind of racial injunction to govern the slowly unfolding migration crisis *before* it overwhelms humanism. It tells us that to persist, white humanism must take hold of the event of "internal climate migration" *prior to* its unfolding.

Why should we take the concept of white affect seriously? What does it offer an analysis of climate change and race? White affect is a useful concept for three main reasons. First and foremost, it allows us to get a handle on the ontology of whiteness in the wider scientific and policy context of climate change. Whiteness, we should remember, is a power that "apportion[s] and delimit[s] which members of the *Homo sapiens* species can lay claim to full human status."[83] Given that whiteness will hardly ever announce its presence in these terms, white affect is valuable because it names the presence of whiteness in our world-historical context in which racisms either pass unnoticed or else come to be expressed in terms of their opposite, "not racism." Naming white affect is especially important, I would add, in discussions of climate change governance where meaningful discussions about whiteness and race are scarce. As Goldberg argues, racism today "cannot be named because nothing abounds with which to name it . . . no terms by which it could be recognized: no precedent, no intent, no pattern, no institutional explication. . . . It is perfectly transparent—a virtual—racism, unseen because see-through."[84] White affect is a valuable term because it names the invisible, yet powerful shared sensibility, the "sounding right," of climate change and migration discourse.

We might be tempted to describe white affect as an artefact of biopower.[85] However, it owes more to what Brian Massumi calls "ontopower." For Massumi, ontopower is concerned with "the regulation of effects, rather than of causes," but the regulation of effects prior to their full actualisation. It is an immanent power that both produces and acts on the emergent event. As a sense-making form of power, ontopower's object-referent is not, however, the physical object—a migrant worker, for example—only the felt sense of the object's presence. Scenarios are crucial to ontopower. They premediate the future and, thus, serve to regulate the effect (climate change–induced migration) before it arises. If climate-migration scenarios make sense this is precisely because they *make sense*.

This brings us to the second reason we should take white affect and its production seriously. Naming white affect will allow us to zero in on how racism is not simply a claim on the future but an investment in making the future appear a certain way. The future is an invaluable political resource. But, as Mike Hulme argues, when the future is reduced to a set of "known" variables, such as the impacts of climate change on demographics or migration, the risk is that "the possibilities of human agency are relegated to footnotes, the changing cultural norms and practices made invisible, the creative potential of the human imagination ignored." When the future is said to be a known entity rather than a plane of immanence, it is robbed of its power to inspire political transformation. A monochromatic future stands in for an irreducible future, governance for immanence, whiteness for the pluriverse. The risk, here, is that a reductive understanding of the future surfaces as a simple, uncomplicated story with which to restore a sense of what it means to be human amidst the fantasy of its undoing. And this, I would argue, is precisely why white affect is so important. It allows for climate reductionism to be conceptualised as an historically specific technology of racist sense-making.

And this brings us to the final reason why naming white affect is important. If white affect is an artefact of ontopower, much of its power resides in the way it invites us to forget the past. And this is more than a little worrying because racisms flourish in the space opened up when our political present is radically severed from any meaningful continuity with the history of race and racism. In fact, we might even say that one of the primary ways racism flourishes today is by intensifying white affect, exchanging the racist past for the immediacy of the future. Naming white affect allows us to appreciate, in turn, how the circulation of racism today is made possible by the sense-making of speculative epistemology.

The significance of white affect is best illustrated, perhaps, with an example from the film *Climate Refugees*. One of the most remarkable features of this film occurs when Newt Gingrich, Nancy Pelosi and John Kerry all declare their shared concern for climate refugees. In a political culture marked by its irreparable fissures, the shared sense among these politicians that climate change is a mounting refugee crisis presages a political consensus sutured by the other of climate change. This is the formidable if not disturbing power of white affect. *Climate Refugees* says nothing about governing migration. All it says is that America's future resides in putting aside old political difference in a renewed commitment to redefining the universal power (i.e., whiteness) to delimit those unfit to be called human. This is the timeless mechanism of human renewal. These old political adversaries may not agree on much. What they agree on, though, is that the other of climate change is *the* difference that will restore American exceptionalism in the face of its undoing.

## NOTES

1. "Postcards from the Future," GMJ, accessed 15 November 2021, http://www.postcardsfromthefuture.com/.
2. Martin Mahony, "Picturing the Future-Conditional: Montage and the Global Geographies of Climate Change," *Geo: Geography and Environment* 3, no. 2 (2016): e0009.
3. Ibid., 14.
4. Ibid., 11.
5. On the topic of anticipatory knowledge, see, for example, Marieke De Goede and Sam Randalls, "Precaution, Preemption: Arts and Technologies of the Actionable Future," *Environment and Planning D: Society and Space* 27, no. 5 (2009): 859–78; Ben Anderson, "Preemption, Precaution, Preparedness: Anticipatory Action and Future Geographies," *Progress in Human Geography* 34, no. 6 (2010): 777–98.
6. Anderson, "Preemption," 778.
7. On "whiteness" as an orientation, see Sara Ahmed, "A Phenomenology of Whiteness," *Feminist Theory*, 8, no. 2 (2007): 149–68.

8. I take the concept of "operative logic" from Brian Massumi who uses it to describe how transhistorical tendencies "are in and of themselves speculatively pragmatic forces." I would argue that anti-Blackness is one such transhistorical tendency and should therefore be approached as a form of pre-emptive power. See Brian Massumi, *Ontopower: War, Powers, the State of Perception* (Durham, NC, and London: Duke University Press, 2015), 3–17.

9. Ash Amin, "The Remainders of Race," *Theory, Culture, and Society* 27, no. 1 (March 2010): 1–23.

10. Derek Hook, *Foucault, Psychology, and the Analytics of Power* (Basingstoke: Palgrave Macmillan, 2007).

11. On this point, one of the clear outcomes of COP26, but also a consistent message over the past twenty-five years, is that governments remain reliably steadfast in their refusal to finance the transition away from fossil fuels even while promising the opposite. The solution to the climate crisis now seems to rest solely with the private sector. See Adam Tooze, "The COP26 Message? We Are Trusting Big Business Not States to Fix the Climate Crisis," *The Guardian*, 16 November 2021, https://www.theguardian.com/commentisfree/2021/nov/16/cop-26-big-business-climate-crisis-neoliberal.

12. Erik Swyngedouw, "Antimonies of the Postpolitical City: In Search of a Democratic Politics of Environmental Production," *International Journal of Urban and Regional Research* 33, no. 3 (2009): 601–20; Bettini, "(In)Convenient Convergences."

13. Daniel Faber and Christina Schlegel, "Give Me Shelter from the Storm: Framing the Climate Refugee Crisis in the Context of Neoliberal Capitalism," *Capitalism Nature Socialism* 28, no. 3 (2017): 1–17; Methmann and Oels, "Fearing to Empowering"; Bettini, "(In)Convenient Convergences"; Sanjay Chaturvedi and Timothy Doyle, *Climate Terror: A Critical Geopolitics of Climate Change* (Basingstoke: Palgrave Macmillan, 2015).

14. Wilderson, *Red, White, and Black*; Wilderson, *Afropessimism*.

15. Wilderson, *Red, White, and Black*, 20–21.

16. My point here is emphatically *not* that those defined as would-be climate migrants or climate refugees are in fact slaves. This is clearly not the case. Rather, my point is that the figure of the climate migrant/refugee occupies the same structural position as the slave in the political ontology of humanism. It is the codification of the figure as living death and nothingness that serves as the foundational condition *for* humanism in order for the human to adapt to climate change as human.

17. Stefano Harney and Fred Moten, *The Undercommons: Fugitive Planning and Black Study* (Brooklyn, NY: Minor Compositions), 53. Emphasis in original.

18. Michel Foucault, "Confessions of the Flesh," in *Power/Knowledge: Selected Interviews and Other Writings, 1972–1977*, ed. Colin Gordon (New York: Pantheon Books, 1980), 194.

19. Hook, *Foucault*, 221.

20. Bettini, "Barbarians."

21. Ionesco et al., *Atlas*.

22. Campbell et al., *The Age of Consequences*.

23. Rigaud et al., *Groundswell*.

24. Brian Massumi, *Parables for the Virtual: Movement, Affect, Sensation* (Durham, NC, and London: Duke University Press, 2002), 35. Emphasis in original.

25. Arun Saldanha, *Psychedelic White: Goa Trace and the Viscosity of Race* (Minneapolis: University of Minnesota Press, 2007), 8. My emphasis.

26. Linda Alcoff, *The Future of Whiteness* (London: Polity Press 2015).

27. Sara Ahmed approaches whiteness as an orientation. My use of whiteness as the virtual as a point of view is similar. See Sara Ahmed, "A Phenomenology of Ehiteness," *Feminist Theory* 8, no. 2 (2007).

28. In this way, my use of affect here follows along lines suggested by Jane Bennett in her book *Vibrant Matter: A Political Ecology of Things* (Durham, NC: Duke University Press, 2010). There, following Spinzoa, Bennett describes affect as referring to "the capacity of any body for activity or responsiveness." Affect is, thus, intrinsically material in its immateriality.

29. Ben Anderson, "Affect and Biopower: Towards a Politics of Life," *Transactions of the Institute of British Geographers* 37, no. 1 (2012).

30. Marieke de Goede, Stephanie Simon and Marijn Hoijtink, "Performing Preemption," *Security Dialogue* 45, no. 5 (2014); Marieke De Goede, "Beyond Risk: Premediation and the Post 9/11 Security Imagination," *Security Dialogue* 39, no. 2–3 (2008).

31. Melinda Cooper, "Turbulent Worlds: Financial Markets and Environmental Crisis," *Theory, Culture and Society* 27, no. 2–3 (2010).

32. Lauren Rickards, John Wiseman and Taegen Edwards, "The Problem of Fit: Scenario Planning and Climate Change Adaptation in the Public Sector," *Environment and Planning C: Government and Policy* 32, no. 4 (2014).

33. Cooper, "Turbulent Worlds."

34. Rickards et al., "The Problem of Fit"; Cooper, "Turbulent Worlds."

35. Cooper, "Turbulent Worlds," 171.

36. Matt McDonald, "Discourses of Climate Security," *Political Geography* 33 (March 2013); Andrew Telford, "A Climate Terrorism Assemblage? Exploring the Politics of Climate Change-Terrorism-Radicalisation Relations," *Political Geography* 79 (May 2020):102–50; White, *Climate Change and Migration*; Emily Gilbert, "The Militarization of Climate Change," *ACME* 11, no. 1 (2012).

37. Schwartz and Randall, *An Abrupt Climate Change*.

38. De Goede et al., "Performing Preemption."

39. Schwartz and Randall, *An Abrupt Climate Change*, 22.

40. Ibid., 7. My emphasis.

41. Wilderson, *Afropessimism*, 249.

42. United Nations Security Council, *Letter Dated 5 April from the Permanent Representative of the United Kingdom of Great Britain and Northern Ireland to the United Nations Addressed to the President of the Security Council*, S/2007/186 (New York: United Nations Security Council, 2007); Stern, *Stern Review*.

43. Campbell et al., "Consequences," 59.

44. Ibid., 59.

45. Jared Sexton, "On Black Negativity, or the Affirmation of Nothing," interview by Daniel Barber, *Society and Space*, 18 September 2017.

46. Wilderson, *Afropessimism*, 250

47. Stefano Harney and Fred Moten, *The Undercommons: Fugitive Planning and Black Study* (Brooklyn, NY: Minor Compositions, 2013).

48. Telford, "Climate Terrorism Assemblage," 1.

49. Ibid.

50. Sharpe, *In the Wake*, 15–16.

51. Still, Sharpe (ibid.) does write about "weather," although not in direct correspondence with climate change.

52. Chakrabarty, "Postcolonial Studies."

53. Anderson, "Preemption, Precaution, Preparedness."

54. Rigaud et al., *Groundswell*. Graphic reprinted under Creative Commons Attribution Licence: CC BY 3.0 IGO

55. Mark Duffield, *Development, Security, and Unending War: Governing the World of Peoples* (London: Polity Press, 2007).

56. I take this reading of performativity from Derek Gregory. See, for example, Gregory, *The Colonial Present*. But also Judith Butler, *Gender Trouble: Feminism and the Subversion of Identity* (New York: Routledge, 1990) and *Bodies That Matter: On the Discursive Limits of "Sex"* (New York: Routledge, 1993).

57. I take this reading of Blackness from Warren, *Ontological Terror*, for whom Blackness stands for, and occupies the place of, nothingness in relation to metaphysics. I do not mean to imply, however, that being Black, or that Black life or experience, is somehow devoid of meaning or is meaningless.

58. Louise Amoore, "Security and the Incalculable," *Security Dialogue* 45, no. 5 (October 2014).

59. Robtel Neajai Pailey, "Decentring the White Gaze of Development," *Development and Change* 51, no. 3 (May 2020).

60. Uma Kothari, "An Agenda for Thinking about 'Race' in Development," *Progress in Development Studies* 6, no. 1 (2006); Pailey, "Decentring"; Kalpana Wilson, "'Race,' Gender and Neoliberalism: Changing Visual Representations in Development," *Third World Quarterly* 32, no. 2 (2011); Sarah White, "Thinking Race, Thinking Development," *Third World Quarterly* 23, no. 3 (2002); Mark Duffield, "Racism, Migration and Development: The Foundations of Planetary Order," *Progress in Development Studies* 6, no. 1 (January 2006).

61. On the double meaning of the concept of *in-formation*, see Michael Dillon and Julian Reid, *The Liberal Way of War: Killing to Make Life Live* (London: Routledge, 2009).

62. Rigaud et al., *Groundswell*, xxv. My addition.

63. Ibid., xxiv.

64. Tasneem Siddiqui, Mohammad Towheedul Islam and Zohra Akhter, *National Strategy on the Management of Disaster and Climate Induced Internal Displacement* (Dhaka, Bangladesh: Ministry of Disaster Management and Relief, 2015), 12.

65. Ben Anderson and Rachel Gordon, "Government and (Non)Event: The Promise of Control," *Social and Cultural Geography* 18, no. 2 (2017): 160.

66. Brian Massumi, *Ontopower: War, Powers, the State of Perception* (Durham, NC, and London: Duke University Press, 2015), 208.

67. Richard Grusin, *Premediation: Affect and Mediation after 9/11* (London: Palgrave, 2010).

68. De Goede, "Beyond Risk"; De Goede and Randalls, "Precaution."

69. Grusin, *Premediation*, 46.

70. Ibid., 47.

71. De Goede and Randalls, "Precaution"; Anderson, "Preemption, Precaution, Preparedness."

72. Grusin, *Premediation*, 50.

73. Ibid., 48. My emphasis.

74. Massumi, *Parables for the Virtual: Movement, Affect, Sensation* (Durham, NC, and London: Duke University Press, 2002), 6–13.

75. Ibid.

76. Back-formation is a tricky concept. I use it to describe how pure potential is slowed down to the point of stability. Back-formation is what happens when pure potential encounters a moment of stoppage.

77. Lake and Reynolds, *Drawing*, 2.

78. McClintock, *Imperial Leather*, 26.

79. David Roediger, *The Wages of Whiteness: Race and the Making of the American Working Class* (London and New York: Verso, 1991); Ruth Frankenberg, *White Women, Race Matters: The Social Construction of Whiteness* (Minneapolis: University of Minnesota Press, 1993); Alastair Bonnett, "Geography, 'Race,' and Whiteness: Invisible Traditions and Current Challenges," *Area* 29, no. 3 (1997); Richard Dyer, *White* (London and New York: Routledge, 1997); Owen Dwyer and John Paul Jones III, "White Socio-Spatial Epistemology," *Social and Cultural Geography* 1, no. 2 (2000); Audrey Kobayashi, "The Construction of Geographical Knowledge—Racialization, Spatialization," in *Handbook of Cultural Geography*, ed. Kay Anderson, Mona Domosh, Steve Pile and Nigel Thrift (Thousand Oaks, CA: Sage, 2003); Ahmed, "Phenomenology"; Kobayashi and Peake, "Racism Out of Place."

80. Derek Hook, "Affecting Whiteness: Racism as Technology of Affect," London: *LSE Online* (2005).

81. Ibid.

82. Ibid. What I am calling the "ontogenetic dynamic" that produces white affect bears strong conceptual and analytical resonance with Sylvia Wynter's reading of Frantz Fanon's concept of sociogeny. Drawing out the full scope of this resonance lies well beyond the scope of the present chapter but is, nonetheless, a study that needs to be done. This is especially the case since Wynter's reading of sociogeny deals directly with the question of adaptation. The fact that I have not taken this up here reflects the fact that I developed the arguments in this chapter prior to realising the applicability of Wynter's reading of sociogeny to my argument. It also reflects the fact that the main conceptual vocabulary I use in this chapter—Massumi's concepts of ontogenesis and affect—draws from a tradition of thought (e.g., Deleuze, Spinoza and Nietzsche) that does not deal with questions of race or Blackness nor with Fanon and Wynter. See Sylvia Wynter, "Towards the Sociogenic Principle: Fanon, Identity, the Puzzle of Conscious Experience, and What It Is Like to Be 'Black,'" *National Identities and Sociopolitical Changes in Latin America* (2001); David Marriot, "Inventions

of Existence: Sylvia Wynter, Frantz Fanon, Sociogeny and the 'Damned,'" *CR: The New Centennial Review* 11, no. 3 (Winter 2011); Frantz Fanon, *Black Skin, White Masks*, trans. Charles L. Markmann (London: Pluto Press, 2017; Paris: Editions de Seuil, 1952).

83. Weheliye, *Habeas Viscus*, 19.

84. Goldberg, "Threat," 23.

85. Anderson, "Affect and Biopower."

## *Chapter 6*

# Adaptive Migration and Racial Capitalism

In September 2009, the United Nations University (UNU) hosted a conference at the Simon Wiesenthal Center/Museum of Tolerance in New York City on the topic of climate change, human migration, and risk. Titled *Insure Me*, the event assembled an unlikely mix of experts from the fields of international migration law, climate change adaptation policy, global finance, and risk management.[1] Whereas prior to this, "climate change and migration" had been problematised mostly in the language of war, security and humanitarian crisis, *Insure Me* marked a turning point in the evolving international discourse on climate change and migration in which a new truth about climate change and migration was starting to take shape. No longer was the figure of the climate migrant/refugee to be thought of as a catalyst for insecurity. Henceforth, it was to be grasped as a self-possessed agent whose migration is said to hold an important key to climate change adaptation. If the old narrative figured the climate migrant/refugee as a failure to adapt, *Insure Me* would mark a paradigm shift in which migration was to be increasingly imagined within international climate change discourse as a legitimate adaptive response to climate change.

Among those in attendance at the UNU event were representatives from the Munich Re Foundation, the philanthropic arm of one of the world's largest reinsurers, and Caribbean Risk Managers Ltd., a small regional insurer, but a subsidiary of Arthur J. Gallagher & Co., one of the world's largest insurance brokerages. That such prominent global insurers would figure centrally in the *Insure Me* event raises important questions about the relationship between global finance, climate change, migration, and risk, just as it raises important questions about how these phenomena converge with race and the political ontology of anti-Blackness. Why would global insurance capital show an interest in migration and climate change? What is at stake when the figure of the climate migrant/refugee is reconceptualised in terms of adaptation and resilience? What can the convergence of insurance and migration tell us about the adaptations of both white humanism and the political? These are just some of the questions I pursue in this chapter.

My aim in what follows is to critically elaborate the logic of *adaptive migration* that would come to international prominence around the time of the *Insure Me* conference in the late 2000s.[2] The new logic of adaptive migration that would take hold around that time is important because it seemed to mark a rejection of the racist geopolitical imagination which up until that point had organised much of the popular discourse on climate change and migration. Like the refusal of determinism discussed in chapter 4, it is almost as though the logic of adaptive migration would supply the discourse with an exit strategy from its whiteness. By recasting migration as a legitimate adaptive response to climate change, as opposed to a catalyst for violence and crisis, this new logic would emphasise not only the positive contribution migration has made to the contemporary world order but also that humanity writ large owes its existence to migration. Human beings, it said, are a migratory species, having always used migration over the course of our collective history as a strategy for adapting to environmental change. Whereas the old narrative had constructed the other of climate change in the familiar racist tropes of threat and victimhood, this new narrative felt more humane, more tolerant of migration and less racist. And yet my aim in this chapter is to suggest that far more is at stake in the logic of adaptive migration than a simple turn to liberal humanism.

In what follows, I examine the seemingly novel and apparently non-racist discourse of adaptive migration using the concept of racial capitalism. Racial capitalism is the subject of a growing body of scholarship that acknowledges the deep and historical imbrications of racism and capitalism.[3] In this chapter, I use it as a framework for theorising the logic of adaptive migration, which had begun to surface as a distinctive discourse around the time of the *Insure Me* event. Many early commentators were quick to explain how this new discourse amounted to a neoliberal political strategy for governing the world's poor in the context of climate change.[4] What these important, early critiques overlooked, however, is the way the logic of adaptive migration is shaped by race, the way it could be understood as a new logic of racial rule. The concept of racial capitalism, by contrast, will allow us to extend and deepen this important set of critiques. It will allow us to see how the logic of adaptive migration does not simply conform to neoliberal principles but how, in doing so, it reworks both race and racism. The chapter builds, in particular, on Arun Kundnani's insight that neoliberal political economy is a "new iteration" of racial capitalism. At the crux of Kundnani's analysis is the following idea that:

> race is not just a means of dividing waged working classes but also the means by which capitalism codes and manages the contradictions between waged and unwaged, between possessors and dispossessed, between citizens endowed with liberal rights and "unfree" labouring populations—from the enslaved to the undocumented. This means that racism is not reducible to a legacy of the past but is *continuously regenerated in new forms* out of globally dispersed divisions of labour and the struggles against them.[5]

The argument I develop in this chapter is that, following Kundnani's interpretation of race and racism, the logic of adaptive migration, while having the appearance of displacing race and racism, in fact, entails the reworking of both. It does this by codifying the very same contradictions between waged and unwaged, possessors and dispossessed, or in the language Mann and Wainwright use to conceptualise the political, the dominant and dominated. Yet it does so in a very precise way and in a manner that differs fundamentally from that of the figure of the climate migrant/refugee. In fact, here, it is important to stress that the figure of the climate migrant/refugee and the adaptive migrant are not the same. Both stand as instances of the other of climate change, but they authorise very different political formations. The former, as I have argued in the preceding chapters, is a signifier of Blackness, of rupture, the undoing of racial nomos, whereas the adaptive migrant stands for the very continuity of racial capitalism, albeit in adapted form. In this way, adaptive migration should be approached as a very particular governing logic; it seeks to govern the *risk of migration* or what we might call the *threat of Blackness*, by converting the risk or threat into a use-value for capital. In this way, adaptive migration is quite specific. Not only is it used to govern neoliberalism's spectacular failure to universalise waged labour. But, as a logic of adaptation, it also serves the additional purpose of supplying racial capitalism with an important means for adapting to the coming geohistorical event of climate change. It reworks what will have been a transformational event—the undoing of racial nomos, threat of Blackness—by seeking to make it *counter-transformational*. That is, the logic of adaptation migration is an attempt to capture what it claims will be a future event of violence and crisis and redirects it towards the ends of securing racial capitalism.

## ADAPTIVE MIGRATION: THE NEOLIBERAL TURN

What is the logic of adaptive migration? And why is it important? We first encountered the logic of adaptive migration in chapter 4. It refers to the seemingly innocuous idea that migration ought to be considered a legitimate adaptive response to climate change rather than a failure to adapt.[6] The logic of adaptive migration also coincides with the explicit move away from determinist reasoning and the commensurate turn to "complex" explanation. It would hardly be an exaggeration to say that this idea is now hegemonic within the international discourse on climate change and migration. According to the logic of adaptive migration, migration in the context of climate change is no longer considered to be inherently problematic but is instead understood to form a part of the solution to the climate crisis. One key feature of the migration-as-adaptation thesis is the way it constructs the figure of the climate migrant/refugee in the language of resilience.[7] By this view, migration, whether seasonal, rural-to-urban, long-term, short-term, domestic or transnational, allows households to diversify their sources of income, thereby expanding their adaptive capacity and making them more resilient to environmental change. This, it goes on to say, is especially important for households located in places heavily exposed to climate risk, including those where economic activity is highly weather-dependent, such as agriculture and tourism, or urban informal settlements. By providing such households with diverse sources of income, migration, it is argued, will improve the capacity of individuals, households and communities to adapt to climate change. On this point, the logic of adaptive migration shares much in common with the new economics of labour migration.[8] But it also reflects what David Theo Goldberg refers to as racial neoliberalism, which is broadly the idea that neoliberalism, as a geographically differentiated modality of political rule, emerged in the 1970s as a means of governing what he calls "the threat of race."[9]

Although it is difficult to pinpoint precisely when the migration-as-adaptation thesis first emerged in the international discourse on climate change and migration, it gathered pace in 2009 in the immediate aftermath of the 2008 global financial crisis. Events like *Insure Me* played their part in consolidating this new discourse, as did a slew of international policy texts. The thesis found early expression, for example, in a 2009 report called *In Search of Shelter*, which contained the findings of a European Commission–funded research programme called Environmental Change and Forced Migration.[10] It was also evident in the International Organization for Migration's anthology *Migration, Environment, and Climate Change: Assessing the Evidence* also published in 2009.[11] Romain Felli traces the thesis's emergence even earlier to the 2007 meeting of the International Organization for Migration Council, as well to the Climate Change, Environment, and Migration Alliance set up a year later by the IOM, UNEP, UNU and the Munich Re Foundation to "mainstream environmental and climate change considerations into migration management policies and practices, and to bring migration issues into global environmental and climate change discourse."[12] While, for Sarah Nash, by roughly the mid-to-late 2000s, a growing and powerful epistemic community was beginning to take shape that actively sought to institutionalise the figure of the climate migrant/refugee in international discourse in the anodyne language of adaptation.[13] As already noted, the thesis expresses a tolerant attitude towards migration and, thus, works as a counter-narrative to the anti-immigration discourses that proliferate so unashamedly today. Refusing the language of migrant exceptionalism, it frames migration as an ordinary aspect of contemporary political economy which can be marshalled for adaptive purposes.

How might we explain this shift in the discourse on climate change and migration? Several commentators have sought to account for it by showing how the seemingly benign idea that migration is a legitimate adaptive response to climate change is, in fact, an innovation of neoliberal capital. Romain Felli argues, for example, that the migration-as-adaptation thesis "follows the nature of neoliberal capitalism in that it produces and reproduces 'primitive accumulation.'"[14] Primitive accumulation is here understood not simply in the way Marx originally used the term, as a means by which people are dispossessed of their land and customary social relations and, thereby, converted into wage labour suitable for capitalist production. For Felli, primitive accumulation is also an ongoing process in which states and international agencies seek to maintain contemporary capitalist social relations through a concerted effort to manage global labour migration. Central to this process is a discourse in which migrant labourers are constructed not as political actors but as individual economic agents. What Felli observes is that the migration-as-adaptation narrative marks a decisive shift from a juridico-political discourse in which climate *refugees* are said to be in need of legal recognition to an economic discourse in which so-called climate *migrants* are construed as economic agents responsible for managing their own vulnerability to climate change. The language of "climate refugees" carries the connotation that those who would be displaced by climate change should be owed some measure of compensation, such as legal recognition or reparations, from those most responsible for climate change (i.e., northern industrial economies). In stark contrast, the migration-as-adaptation thesis places responsibility for adapting to climate change not on states but on so-called climate migrants themselves. No wonder Western governments and international agencies would come to embrace this new logic. Not only does it relieve industrial economies from the moral burdens of culpability and compensation. It also places moral responsibility to adapt to climate change squarely onto those workers who are said to be most vulnerable to climate change.

Thus, as Felli observes, the migration-as-adaptation thesis promotes labour migration as a means for managing climate vulnerability. Migrant workers are encouraged to sell their labour on the global labour market in exchange for wages which they can then remit back to their home communities as important sources of investment for building resilience. In a world context in which Western governments adamantly refuse to compensate those adversely affected by climate change, it is easy to see why the migration-as-adaptation thesis has become so popular. If, in the language of insurance capital, climate change can be understood as an unfunded liability, then the logic of adaptive migration promises to harness the vast global remittance economy to meet the financial requirements for adapting to planetary climate risk. It is also easy to understand why international agencies would tout migration management schemes, such as temporary and circular migration, as practical examples of adaptive migration. By redirecting workers from the Global South to regions and sectors of the global economy where it is needed the most, such schemes could be used to satisfy the labour requirements necessary to ensure that global circuits of capital can themselves adapt to climate change.

Chris Methmann and Angela Oels offer a slightly different explanation. Whereas Felli insists that the logic of migration-as-adaptation reflects neoliberal economics, Methmann and Oels argue that the thesis is political. But not political, as in the demand for recognition or rights, but political as a logic of rule.[15] Their claim is that resilience is a form of neoliberal governmentality, promoted by international institutions to govern the contingent relation between climate change and migration. By governmentality they refer to the concept made famous by Michel Foucault who used it to explain why the state came to be concerned with the act of governing of populations. Specifically, they argue that resilience is a technique of advanced liberal government used to govern populations through contingent events, such as disasters or emergencies. Resilient populations are those said to be best equipped to accommodate the disruptions produced by such events. Abrupt climate change is their case in point. If improperly managed, the social disruptions of abrupt climate change could result in societal and political breakdown, symbolised all too often by the figure of the climate migrant/refugee. For Methmann and Oels, the migration-as-adaption thesis is a form of political rule which encourages populations to become resilient by using migration to live through, or adapt to, the contingent event of climate change. Whereas Felli claims the thesis is primarily about disciplining surplus labour in the interest of securing capitalism, Methmann and Oels argue that the thesis promotes resilience as a distinctive form of neoliberal government aimed at securing the living conditions and lives of those affected by contingent events.

Giovanni Bettini reaches similar conclusions about the thesis's relation to neoliberalism, and, like Methmann and Oels, draws attention to another important facet of the migration-as-adaptation thesis which is that any form of adaptive migration will always imply its opposite, maladaptive migration.[16] Thus, Bettini distinguishes between what he calls "good circulation" and "bad circulation," adaptive migrants who promote resilience through their participation in remittance economies (good circulation) and those who do not (bad circulation), for example, those who seek asylum in Europe or the so-called trapped populations that wind up settling in areas of high environmental risk, such as urban informal settlements.[17] For Bettini, this suggests that the migration-as-adaptation thesis is really just a form of biopolitical security in which liberal states promote desirable forms of migration, while intensively policing or abandoning those whose migrations are said to fall outside the parameters of acceptability.[18] Notwithstanding the important differences that separate these four authors, together they provide a convincing argument that the migration-as-adaptation thesis arose as a neoliberal response to the threat the migration effects of climate change would pose to capitalist political economy.

We could even double down on these important arguments by showing how the logic of adaptive migration resolves the definitional dilemma at the heart of climate change and migration discourse in a manner consistent with neoliberalism. Consider, here, how the figure of the climate migrant/refugee confounds the decisionism of political sovereignty. If the figure of the climate migrant/refugee is often said to pose a threat to state stability, it follows that any attempt to ameliorate such instability on the part of the state would require policing or controlling climate migrants or climate refugees. The problem, however, is that any such effort to control so-called climate migrants/refugees would require a sovereign agent, such as the state, the law or a delegated international institution having to decide who is and who is not a climate migrant or refugee. In other words, the sovereign would be faced with having to distinguish the exceptional from the ordinary. A similar political decision would also be required if the figure is said to need special legal recognition, for example, as a precondition for accessing humanitarian assistance or some other form of compensation. Thus, from the perspective of political sovereignty, the fundamental questions that arise within the discourse on climate change and migration are: Who is a climate migrant/refugee? How is the figure defined? How many are there? And how can they be identified and controlled? Indeed, inherent to these questions is the logic of the exception, which is said to reside at the core of any conception of political sovereignty.[19] Yet, as we saw from chapter 1, any such sovereign decision would be completely arbitrary, since migration is multicausal and, therefore, irreducible to climate change alone. Too many other variables (e.g., land tenure, labour markets and political violence) can also be used to explain migration, which together make it impossible to differentiate so-called climate migrant/refugees from other migrants. In fact, even if the sovereign were compelled to make such a decision, it would be unable to reach one because there would be no decision to make. The figure, as we saw in chapter 2, is ultimately undecidable. And yet, paradoxically, because the relation between migration and climate change is said to be a pending problem for the state (i.e., it threatens security, humanitarian crises and the like) a decision needs to be made; the climate migrant/refugee needs to be identified and controlled in order to avoid the calamitous future foretold by the political logic of sovereignty. The state can't decide, but it must decide. This irresolvable paradox remains one of the distinctive features of the political and legal ontology of climate change.

The migration-as-adaptation thesis, by contrast, resolves this fundamental quandary by reformulating the sovereign decision into a question about one's exposure to risk. Thus, rather than posing the sovereign question, namely, how are we to define the category "climate migrant," which is an unanswerable question, the migration-as-adaptation thesis instead asks: Who is at risk of migration due to climate change? How do households use migration to adapt to environmental risks? How can adaptive migration be managed? In other words, the migration-as-adaptation thesis moves the political decision away from the binary logic of the sovereign state, which seeks to distinguish inside from outside, ordinary from exceptional, citizen and non-citizen, to the logic of neoliberalism which distinguishes and, thus, revalues people on the basis of their exposure to risk. Indeed, if one of the hallmarks of neoliberal rationality is that any decision taken by the sovereign state is always bound to fail (i.e., "markets know best"), then the migration-as-adaptation thesis mimics this logic to the letter.[20] It says that because no amount of information is available that will help the sovereign identify a climate migrant/refugee, the best we can do is map those at risk of migration and then make sure they have the resources they need to manage those risks. *Groundswell*, the World Bank report discussed in the previous chapter, deploys precisely this kind of reasoning. Through the use of scenarios, it translates climatic uncertainty into the more calculable measure of risk, mapping, in particular, those at risk of internal migration.[21]

There is every reason to take these arguments about neoliberalism and adaptive migration seriously. Not only can they help us grasp how neoliberal power is exercised at the confluence of two of the world's most pressing planetary crises (climate change and migration), so they also remind us that there is nothing neutral or benign about the migration-as-adaptation thesis. These arguments can also help clarify why and how powerful global insurance companies might have a stake in the debate on climate change and migration—adaptive migration is both a form of risk management and a potential source of resilience financing for a world on the verge of climatic breakdown. Adaptive migration, thus, becomes a potentially lucrative vehicle for financing the unfunded liabilities of climate change in the Global South. Thus, where Joel Wainwright and Geoff Mann argue for "a robust political language defending the right of people to migrate in anticipation of climate change," Bettini's arguments suggest they should probably look somewhere other than adaptive migration.[22] As an expression of neoliberal rule, adaptive migration, Bettini tells us, is "not necessarily conducive to a progressive turn, able to support the rights of prospective (climate) migrants."[23]

For all these reasons I remain strongly sympathetic with the argument that adaptive migration is coterminous with the rationality of neoliberalism. Yet even while I find it convincing and indispensable, I also find it lacking in three key respects. First and foremost, this interpretation of adaptive migration as a neoliberal technology overlooks how geopolitical discussions about migration and capitalism have always been fundamentally shaped by histories of race and racism.[24] By neglecting this important fact, these critics are unable to see how the logic of adaptive migration also involves the rewriting of race and racism. Second, I wonder if these critiques are too focused on theorising the logic of adaptive migration as a technology for governing the movements of actual migrants, as opposed to the one more focussed on converting the *risk of migration* into a use-value for capital. And finally, I wonder whether this argument obscures something even more fundamental for how we might grasp the relationship between climate change, migration, racial futurism and capitalist political economy. Borrowing from Sara Nelson and Bruce Braun, I wonder whether this line of argument is too quick in drawing an inevitable, or necessary, connection between adaptive migration and neoliberal rule and, therefore, too quick in dismissing adaptive migration as a potentially useful political concept for developing alternative responses to climate change.[25] Might there be other ways of thinking about adaptive migration and neoliberalism? Bettini hints at this possibility in the above quotation. Would it be more advantageous, perhaps, to think of this connection as contingent rather than necessary? Maybe the problem is not adaptive migration *per se*, but the way it seems to get absorbed into the political economic rationality of neoliberalism? Adaptive migration is, after all, a feature of human existence that long predates anything called neoliberalism.

In the remainder of this chapter, I reinterpret the logic of adaptive migration using the concept of racial capitalism. I suggest how racial capitalism has the advantage of addressing these shortcomings, while, at the same, allowing us to understand how the logic of adaptive migration is more than just a neoliberal technology, but also an important site for understanding how race and racism are being reworked and, thus, redefined, as a condition for the adaptation of capitalism itself.

## RACIAL CAPITALISM

What is racial capitalism? Racial capitalism is both an analytical framework and the name given to the concrete material condition of global capitalism. At the core of both conceptions of racial capitalism is the understanding that racism is neither an aberrant phenomenon nor an historical accident, but one of capitalism's defining attributes. Space precludes a full genealogy of the concept of racial capitalism; suffice to say it was first used to characterise the Apartheid system in South Africa and later expanded by Cedric Robinson in his seminal text *Black Marxism: The Making of the Black Radical Tradition.*[26] Racial capitalism has since been used to explain racisms across numerous contexts, including environmental racism.[27] While even more relevant for our purposes, Neel Ahuja has used it to theorise how the political discourse on climate change and migration obscures the historical expansion of global fossil fuel production and its resulting effects on global patterns of racialised migrant labour.[28] We will look at Ahuja's important account of racial capitalism momentarily.

However, it is Arun Kundnani's reading of racial capitalism that is most pertinent for understanding how the neoliberal logic of adaptive migration involves the reworking of race and racism. "Neoliberalism," he argues, "cannot be understood—or politically opposed—without taking into account how it reorganizes and reconstitutes racism to produce a new, integrated structure specific to the historical moment of neoliberalism."[29] From this perspective, "racial domination does not simply survive under the seemingly racially neutral auspices of neoliberalism *but is actively reworked* as an internal aspect of [neoliberalism's] redrawing of the social, the political, the cultural and the economic."[30] In other words, race and racism were always a key part of the process by which capital became neoliberal. Or, more generally, capitalism is so reliant on racism that it must continuously reconstitute racism as a condition of its own continuous adaptation. My contention is that the logic of adaptive migration should be understood as the very means by which global racism is being reformulated to suit the needs of neoliberal political economy, as the latter struggles to adapt to climate change. It should be understood to lie at the heart of what Mann and Wainwright call the "adaption of the political."

To build this argument, let us first consider Kundnani's interpretation of the political philosophy of Friedrich Hayek, who is widely considered to be one of the foundational thinkers of contemporary neoliberal capitalism.[31] What Kundnani shows is that race is reworked in Hayek's political philosophy as the ideological means by which the contradictions internal to his philosophy are managed. Broadly speaking, Hayek was convinced that universal human welfare could be achieved by allowing people everywhere to compete economically in a kind of universal market. This would be possible, however, only on the condition people could do so unencumbered by what Hayek considered to be the misplaced, coercive tendencies of the state. Kundnani's interpretive innovation is to show that, while, for Hayek, market competition was assumed to be both natural and universal, its origins are firmly situated, paradoxically, in "the intellectual legacies of western civilisation."[32] This, in turn, creates a fundamental problem at the centre of Hayek's political philosophy. If market competition is a universal logic, it nevertheless remains "haunted by its failure to explain how its implementation could be universalised without undermining its own nominal universal values."[33] In other words, the question Hayek faced was how to square his philosophy of the market as a *universal* institution with those unable or unwilling to conform to the purportedly universal logic of the market. This question would be especially pertinent in relation to the non-Western parts of the world, where the supposedly universal values of the market were not always so self-evident. This is where race enters the picture; for Hayek, the answer to this conundrum was to be explained in terms of race. Although Hayek would reject "the concept of race in any physiological sense," he would explain this contradiction by resorting to what Étienne Balibar has called "neo-racism."[34] This is a type of racism which would emerge after World War II to explain group difference on the basis of culture rather than biology. Thus, as Kundnani explains, for Hayek, those unable to adapt to the universal logic of market competition were simply written off as culturally deficient—something about their culture prevented them from claiming their universal humanism. It was, therefore, down to the purveyors of neoliberalism to defend this most precious universal market order from those they imagined to be culturally predisposed to undermine it. Hence, Hayek would advocate the use of immigration controls and other forms of policing to prevent those racialised as culturally deficient from eroding the social foundations of market order. While other neoliberals, like William Röpke, would defend South African Apartheid, just as Milton Friedman would support white minority rule in Rhodesia.[35] Here, we should understand "race" (now codified in the language of culture) to be the political technology Hayek would use to explain away a contradiction at the heart of his philosophy: the difference between those who would embrace the rationality of the market and those who would not or could not. Thus, as Kundnani puts it, neo-racism "enables neoliberalism to oscillate between thinking the limits it encounters are the necessary consequence of inferior others and thinking they can be overcome by renewed imposition of its market logic."[36]

Kundnani then goes on to explain how race is used to govern a similar anxiety at the heart of actually existing neoliberal political economy. His explanation draws on a structural distinction within the global division of labour between, on the one hand, waged workers (i.e., factory workers, the proletariat) and, on the other, the vast global surplus of unwaged workers essential to social reproduction (i.e., rural peasants, workers in informal economies). This latter segment of workers amounts to a massive global surplus population that sits outside the scope of formalised work in the global economy—"not even able to serve as a reserve army of occasional labour"—while remaining essential to it by providing the basic essentials of social reproduction, such as raising children, affective labour, subsistence agriculture and so on.[37] As Kundnani puts it, "Neoliberalism is haunted by the existence of these surplus populations."[38] But what is the precise nature of this anxiety? Haunted by what exactly? Surplus populations represent not only neoliberalism's absolute "failure to universalise its market order."[39] Even more important, surplus populations represent "a space from within which resistance is generated." That is, such surplus populations are haunting because they represent an alternative to the universal market order espoused by neoliberalism. It is from within such spaces that alternative political imaginaries start to take shape and that pose a fundamental threat to the presumed universality of market logic. They are, in other words, a terrifying prospect for the champions of neoliberalism, who must then codify surplus populations in ways that will rationalise and, thus, allay their anxiety. For Kundnani, race is the political technology that enables this to happen. It allows for the "limits to the universalisation of neoliberalism to be naturalised and dehistoricised."[40] Thus, as Kundnani explains:

> The surplus dispossessed come to be represented through a series of racist figures—"welfare queens," "Muslim extremists," "illegals," "narcos," "super-predators," and so on—as part of the process of securing neoliberalism in the realm of ideology. These figures of economic dependency, violations of property, and threats to western culture rework older legacies of racism to produce images that are distinctive to the neoliberal era.[41]

No doubt we should add the figure of the climate migrant/refugee to Kundnani's list. Although each of these "racist figures" is mobilised differently and for differ racial-political projects, what they share in common is that all are said to pose a threat to neoliberalism, when in actual fact, according to Kundnani, they simply "serve as displaced signifiers of neoliberalism's failure to universalise its legitimacy."[42] Race simply papers over neoliberalism's colossal failure to be taken up universally by everyone everywhere. No wonder, then, some would argue that climate refugees were the underlying cause of the Syrian civil war.[43] Nor should it be surprising that some would characterise the war's actual refugees as "climate refugees." In both cases, doing so masks the cataclysmic failure that neoliberal agricultural policies would have in rural Syria, and which ultimately led to the ensuing civil uprisings and, eventually, the war itself.[44]

What Kundnani is describing *is* racial capitalism; race is precisely what allows capitalism to explain away its own internal contradictions. It is important to recognise, however, that racial capitalism is far from universal in form. As a technology for codifying the division of labour, race is incorporated into the logics of capitalism in markedly different ways across time and space. The racisms of various settler colonialisms are, for example, vastly different from those used in plantation slavery to those that would organise post-imperial urbanism in places like Britain and France. What these geographically dispersed and historically differentiated racisms all have in common, however, is that all use "race" to naturalise, dehistoricise and, thus, explain away a whole range of unequal social relations (e.g., possessed and dispossessed, modern and traditional, waged and unwaged, exploitable and unexploitable, dominant and dominated), relations which elicit anxiety among those who are structurally white, but which are, nevertheless, necessary for the functioning of capitalism. So, when Kundnani explains the "racial constitution of neoliberalism," he is explaining how race organises only a very specific historical formation of capitalism. Still, his analysis zeros in on a specific, possibly even universal, aspect of racial capitalism, which is that the continuous use of race to codify unequal social relations is born out of a perpetual white anxiety that arises as a result of its own glaring contradictions. He shows, for example, how dominant groups, in his case neoliberals, including neoliberal institutions, use the concept of race to mollify their anxiety in relation to those they strive to dominate.

Kundnani's interpretation of race places it firmly at the centre of the psychic structure of capitalism. It says that race is not solely ideological but performs an important psychic function by which dominant groups disavow how they themselves bring about the very conditions of their own undoing, conditions that, in turn, warrant the concept of race. So, for example, as Kundnani puts it, neoliberalism famously invents a set of racist categories—"illegals," "extremists," and so on—onto which neoliberalism displaces *its own* failure to universalise its logic. The same, I would suggest, can be said of the figure of the climate migrant/refugee. White humanism invents this racist category onto which it displaces both its own role in bringing about the crisis of climate change and its own decades-long failure at resolving the crisis.[45] In other words, Kundnani's argument about the racial constitution of neoliberalism is important because it allows us to understand how the figure of the climate migrant/refugee, a racial category *par excellence*, comes to resolve the crisis of whiteness at the heart of climate change. The figure is used to shore up white humanism in the very moment the human is forced to confront its own culpability in bringing about the ontological terror of climate change. Race, in this sense, is precisely the technology which promises to restore the human, indeed, the *white* human, in the face of its own undoing.

Neel Ahuja's interpretation of climate change, migration and racial capitalism adds a layer of historical depth to this argument.[46] For him, the global rhetoric of climate change and migration, symbolised by the figure of the climate migrant/refugee, obscures the historical conditions that produce both climate change *and* the political economic vulnerabilities that put millions around the world at risk of having to migrate because of climate change. A full account of this history lies beyond the scope of this chapter. But at the crux of Ahuja's history are three interrelated phenomena. First, the expansion of fossil fuel extraction by U.S. foreign investment in Saudi Arabia and the Persian Gulf after World War II—a form of extraction itself heavily reliant on a racist division of labour[47]—contributed massively to the rise of global greenhouse emissions worldwide. Second, the combined effects of the 1973 oil crisis, the removal of the U.S. dollar from the gold standard and the huge debt incurred by the U.S. government to fight the Vietnam War all put in place the macro-economic conditions that would eventually bring about the structural indebtedness of dozens of countries in the Global South. These debts would, in turn, push countries across the Global South towards export-led development in order to acquire the foreign exchange (i.e., U.S. dollars) needed to service these very debts. Bangladesh, for example, would transform much of its Southern coast into shrimp farms which would devastate the local environment and render landless large numbers of rural peasants who would be forced to pursue more precarious forms of labour in Dhaka and Chittagong.[48] And third, the oil wealth accumulated in Saudi Arabia and the Gulf States spawned a massive demand for labour that would be supplied in large measure by South Asian workers, including those from Bangladesh. The resulting migration corridors represent a major source of remittance income that South Asian migrant workers are now expected to access for the purpose of fostering resilience in their home countries.

Although much can be said about the specific details of Ahuja's history, it is important for our discussion because it adds historical substance to Kundnani's claim that racial capitalism manages its own contradictions with race. Ahuja's history connects up the geopolitics of oil extraction with the ensuing global hegemony of the U.S. dollar and links these directly to patterns of South Asian transnational migration and to the structural conditions that have produced surplus populations that are now said to be at risk of migration, especially in South Asia. His point, however, is that media and policy narratives about the figure of the climate migrant/refugee obscure this history. As such, those who are now said to be at risk of migration because of climate change are codified not in terms of structural indebtedness, or the geopolitics of the U.S. dollar, or even the history of fossil fuels, but in terms of the geophysical impacts of climate change. Scrubbed clean of the history that shapes their vulnerability, they now serve as a global surplus population onto which Western humanism is able to project its own longing for order in a world on the verge of disorder yet without having to fully confront its own contribution to climate change. In Kundnani's terms, we could say that, within the broad global context in which capitalist political economy is struggling to adapt to climate change, the figure of the climate migrant/refugee *is* race. Euro-western humanism displaces both its contribution to climate change and its colossal failure to mitigate climate change onto the other of climate change. It does this by codifying, or perhaps even better, *naturalising*, the global spatial division of labour into two distinct camps: those who occupy the unmarked position of universal humanism—a benign, purportedly stable Euro-American humanism, for example—and the global surplus population who by virtue of their historically produced precarity are imagined to pose a fundamental threat to humanism, a threat now said to be amplified by climate change. The ensuing racism then polices the distinction between the two. Hence, the border becomes the paradigmatic means by which surplus populations are controlled through brutal forms of racist violence.

Kundnani and Ahuja's conceptualisations of racial capitalism are together indispensable for helping us understand how race structures the discourse on climate change and migration. Both can explain the racism at stake in the figure of the climate migrant/refugee. They are, however, less useful for explaining how the logic of adaptive migration involves the reworking of race and racism. This is partly because both, and quite understandably, place so much emphasis on how the racial state polices surplus populations at the border. Here, the border can be understood in both its literal sense—as a territorial demarcation which gives rise to a range of bordering practices enacted by the racial state either on the border (e.g., walling, border patrols) or beyond the border (e.g., intensive policing of immigrant communities, criminalising undocumented workers, creating a hostile environment)—and in its discursive sense—as the practice of drawing boundaries between people (e.g., us/them, citizen/non-citizen, freedom/unfreedom). Either way, their emphasis on the border/boundary reinforces the idea that racial capitalism is maintained at the border. And while it is almost certainly the case that racial capitalism will seek to adapt to climate change by reinforcing border security—indeed, it already seems to be doing just this[49]—the logic of adaptive migration now institutionalised in the international system suggests that the adaptation of racial capitalism to climate change is more multifaceted than a single-minded reliance on border enforcement. What Kundnani and Ahuja both seem to emphasise is racial capitalism's reliance on political sovereignty (although this is less the case for Ahuja). Whereas the logic of adaptive migration would seem to suggest a different kind of relationship to racial capitalism. Not one resolutely focused on impeding migration, but something different. Not a racism enacted on the border, but a different sort of racism. As we saw earlier, one of the unique features of the logic of adaptive migration is that it resolves the paradox of identifying climate migrants/refugees by reformulating the sovereign decision into a question about one's exposure to risk. The argument I develop in the following section is that the logic of adaptive migration reworks race and racism in the language of risk.

## ADAPTIVE MIGRATION AND THE ADAPTATION OF RACIAL NEOLIBERALISM

How does the logic of adaptive migration rework race and racism? Like the figure of the climate migrant/refugee, the logic of adaptive migration is also racial. It, too, codifies the global division of labour in a way that masks neoliberalism's failure to universalise the wage. It does so, however, in ways that differ fundamentally from the figure of the climate migrant/refugee. The latter is racial because it is undecidable, indeterminate and ambiguous. It is monstrous and scary, and, therefore, something against which humanism must defend itself. In this way, the figure of the climate migrant/refugee authorises the exercise of political sovereignty. Hence, the priority given to policing the border. The adaptive migrant, by contrast, is racial insofar as it designates an idealised form of difference that can be incorporated differentially into racial capitalism. The adaptive migrant is a difference that doesn't so much threaten capitalism, but one that sustains, even nurtures, capitalism, provided it follows a carefully delineated script. The implication being, however, that if the adaptive migrant *doesn't* conform to the script set for it by the terms of the logic, then it *will eventually* pose a threat to capitalism. *The Foresight Report* is clear about this, for example, when it writes:

> Unless government policies are strengthened so that people can proactively and safely use migration as one of a range of adaptation strategies, or take refuge safely in a time of crisis before returning to their home, *more negative outcomes will occur*. Depending upon the scenario, people may be trapped in vulnerable areas or lose their choice to stay in certain areas, *displacement with geopolitical consequences may occur* or migration may be unpredictable and unexpected and concentrated on particular locations.[50]

Thus, according to the *Foresight Report*, ordered and orderly migration carried out under the governing logic of adaptive migration is what is needed to prevent the world from falling into a Hobbesian state of nature. This does not make the figure of the adaptive migrant any less racial than the figure of the climate migrant/refugee. The logic of adaptive migration is simply another means of policing global surplus populations in the context of climate change. However, instead of codifying them racially as barbarous, violent and threatening, the logic of adaptive migration codifies global surplus populations by rendering them useful for capitalism. Indeed, this is precisely Felli's argument.

So how, then, is this logic racial? How does it redefine race? The logic of adaptive migration is racial, first, because it naturalises risk, and then, second, because it converts those who are said to be most at risk of migrating because of climate change into a use-value for capital. It is, in this sense, a technology of power that converts *potential* climate-induced migration, the very threat of Blackness, the very real possibility that millions of the poorest and most vulnerable people around the world will need to relocate because of the climate crisis, into an exploitable opportunity that, if managed properly, stands to fuel a new a round of capital accumulation. From this perspective, the logic of adaptive migration takes what will have been a transformational event in the context of climate change—the undoing of racial nomos—and converts that future event into an exploitable resource. That is, by translating the risk of migration into a use-value for capital, it makes the transformational event of migration *counter*-transformational.

Two concrete examples can be used to illustrate how the logic of adaptive migration reworks race and racism in support of the adaptation of racial capitalism. The first comes from Neel Ahuja's account of migration in and from Bangladesh. The second is taken from an obscure and mostly neglected report published by the Asian Development Bank in 2012 called *Addressing Climate Change and Migration in Asia and the Pacific* (*ACCMAP*).[51]

In the first example, race is reworked in the distinction between adaptive migration and unmanaged, or maladaptive, migration. As discussed above, much of Ahuja's analysis focuses on the way the discourse on climate change and migration obscures the historical production of vulnerability and risk in places like Bangladesh. His point here is simple: one of the primary effects of the discourse is to naturalise risk, to make risk appear as though it were an objective condition. The result is that it directs us to consider how best to foster adaptation so that people have the capacity to overcome, or at least manage, these conditions of risk. Ahuja explains how the Bangladeshi state has identified transnational migration as an important risk management strategy for fostering climate adaptation in Bangladesh. Accessing the global remittance economy through various forms of labour migration thus becomes a primary means the Bangladeshi state uses to attract foreign dollar remittances. These remittances are then, in principle, used to help Bangladesh adapt to climate change. Thus, according to the *Bangladesh Climate Change Strategy and Action Plan* (2009), "Preparation in the meantime will be made to convert this population [what it earlier calls 'environmental refugees'] into trained and useful citizens for any country."[52] This should not be taken as an indictment of the Bangladeshi state; within the global division of labour, it is simply pursuing one of the only strategies available to it for accessing foreign capital. What matters for our discussion, however, is that the logic of adaptive migration winds up naturalising the distinction between those in the Global South who are able to access formalised transnational networks of labour, and those who cannot. It naturalises, in particular, the separation between Bangladeshi transnational labour migrants—those who have managed to acquire the social, economic, cultural, political and financial capital needed to access networks of formalised work (e.g., domestic workers who service middle- and upper-class households in the Gulf)—and those unable to acquire these capitals whose labour (e.g., manual labour, social reproduction, affective labour, subsistence agriculture) falls outside the scope of formalised work, but who nevertheless remain fundamental in order for transnational labour migration to persist.

The logic of adaptive migration is powerful in this example because it invites us to overlook why so many millions of Bangladeshis who fall into the latter group live in conditions of such extreme poverty. The logic is not remotely concerned with the history of inequality in Bangladesh. Nor does it propose to reverse the structural conditions that produce this inequality. (In Ajuha's fascinating account, this history is inextricably bound up with the history and geopolitics of the U.S. dollar, which is itself tethered to the history of fossil fuel.) What matters to the logic of adaptive migration is not why people are vulnerable, but whether and how they use migration to manage their vulnerability. Here, the concept of racial capitalism is invaluable. It allows us to appreciate how race is used to codify inequality in contexts such as this. The logic of adaptive migration is a technology of race, in this sense, because it codifies the difference between those able to access transnational labour migration and those unable to do so. Adaptive migrants are those who are said to possess the capital needed to access transnational labour migration, while migrants who fail to manage their risk properly are said to be maladaptive. An example of this latter group is the "trapped populations" category. Inasmuch as they are said to have failed in managing their risk, trapped populations fall outside the parameters of acceptable migration. They are in this sense an example of what Bettini would call "bad circulation." For our purposes, the distinction between adaptive and maladaptive migration is a racial distinction inasmuch as it naturalises would-be migrants on the basis of their adaptive capacity. It says that some would-be migrants possess this capacity while others do not, yet it offers no systematic explanation to account for this difference. It simply accepts this difference as a natural given, which it then uses to codify global surplus populations as a technique for governing labour migration in the context of climate change.

In the second example, the figure of the adaptive migrant is reformulated into an object of financial speculation. In this example, race is reworked in the distinction between insurable and uninsurable migrants. This example comes from the Asian Development Bank report mentioned above.[53] In many ways, *ACCMAP* is unremarkable. Like so many international development reports on climate change and migration, it embraces the idea that migration should be conceptualised as a legitimate adaptive response to climate change and recommends that national governments endeavour to include migration within their development and adaptation planning. What sets the report apart, however, is that it includes a lengthy discussion on how the costs associated with "climate change–induced migration" can be met by leveraging global capital markets through innovative financial instruments, such as weather derivatives, parametric insurance and catastrophe bonds. In this way, *ACCMAP* reconceptualises the event of "climate-induced migration" as a massive financial liability which, if left unfunded, poses considerable financial costs to various states, international agencies, migrants and places of origin and destination. From this perspective, the threat of Blackness is approached not as a geopolitical, legal or security problem, but one of finance. One simply needs to figure out what the costs are, how they should be met and who should pay.

According to *ACCMAP*, the costs associated with "climate change–induced migration" fall into three categories: *up-front costs*, such as transportation, humanitarian relief, data collection and costs to host communities; *economic losses*, including lost productivity, property loss and economic deterioration at the place of origin; and *intangible costs*, such as mental illness, family separation, loss of culture/language and constrained liberties. It notes, however, that because the impact climate change will have on migration is so uncertain, these costs are themselves also highly uncertain. Amidst such uncertainty how then might these costs be met? Unsurprisingly, some of these, it says, might be met by any number of international agencies, such as the International Organization for Migration. But the report also makes the remarkable claim that global capital markets are well positioned to meet them too. *ACCMAP* falls short of directly advocating the use of insurance and reinsurance to fund migration-related costs. It does, however, openly speculate on how catastrophe bonds and weather derivatives might be used to fund these contingent liabilities. A full description of each instrument lies beyond the scope of my analysis. However, in brief, unlike ordinary insurance instruments, which price risk according to historical frequency and compensate for actual losses, catastrophe bonds and weather derivatives price risk by modelling *future* risk events and then pay compensation should a set of risk parameters be met.[54] Catastrophe bonds and weather derivatives are both tradeable financial assets whose prices are set by global financial markets.

One of the distinctive features of *ACCMAP* is that it reconstructs the figure of the climate migrant/refugee (i.e., the potentiality of Blackness) in the language of financial risk. Or more accurately, it translates the figure's indeterminacy, undecidability and excess into a set of *potential* costs which allow for the figure to be grasped as a future financial liability. Why is this so consequential? Because putting a price on some of the world's most vulnerable people, even in the abstract, is the first step in converting them into a fungible commodity. *ACCMAP* puts this in no uncertain terms when it claims that "to the extent that any aspect of climate-induced migration can be turned into a financial commodity or insurance policy, the market clearing price becomes the ultimate expression of all personal opinions simultaneously, adding objectivity to temper political motivations in debates over climate change."[55] To put this in very straightforward terms, the report is an attempt to conceptualise how the climate risk faced by the Black and brown majority world can be translated into a tradeable financial asset.

Thus, two questions arise. First, how does this financial logic operate in support of adaptive migration? And how can it be understood in terms of racial capitalism? To the first of these questions, catastrophe bonds and weather derivatives are imagined in *ACCMAP* as a means of meeting the financial liabilities associated with adaptive migration. Depending on how these instruments are designed, they can either supply states or intermediaries (e.g., non-governmental organisations) with the short-term liquidity needed to ensure that migration is adequately managed in the context of an environmental event. This liquidity can then be used to police migration to ensure it is "properly" adaptive. Or these instruments can supply individual migrants and households with the liquidity they might require in order to use migration as an adaptive strategy. The latter might include, for example, the costs associated with transportation, food or housing, or even the costs of accessing transnational labour markets. In answer to the second question, these instruments convert global surplus populations, those most at risk of being displaced by climate change, into financial opportunities for global insurance and finance capital. Race is never explicit in this formulation either. Still, these instruments operate racially inasmuch as they displace neoliberalism's failure to universalise wage labour by reformulating some of the world's poorest and most vulnerable to climate change into the categories of insurable and uninsurable. Those said to be insurable are, thus, viewed as opportunities for financial gain, while latter are not. Not only do these instruments codify the other of climate change in terms of naturalised risk, but they also seek to place a formal price on this risk which would then be made to circulate in monetary form. In this way, through these financial risk management tools, it is possible to see how neoliberal political economy undertakes to deepen and extend the logic of the market into the very conditions of survival faced by those most at risk of climate change. It takes those whose lives stand to be most adversely affected by climate change, dispenses with the history of their vulnerability and converts them into objects of financial speculation.

In our postracial condition in which race continues to organise the global division of labour, albeit in the absence of direct references to race, I would suggest that we approach the logic of adaptive migration as a racial technology.[56] From the perspective of racial capitalism, this is because the logic of adaptive migration recodifies neoliberal capitalism's failure to "universalise its market order."[57] Some readers will undoubtedly find this reasoning difficult to accept. After all, adaptive migration is often framed, even if implicitly, as a non-racist alternative to the racial logics of sovereignty, deterministic reasoning and geopolitical security, ordinarily associated with the discourse on climate refugees. But from the perspective of racial capitalism, race and racism are not reducible to the logics of the border. Nor are they akin to xenophobia or an irrational fear of otherness. Rather, race and racism are concepts that capitalism uses to manage surplus populations. The border is simply one mechanism in racial capitalism's arsenal. The logic of adaptive migration, I would suggest, is another. Of course, it will never announce itself as a logic of race. Nor does it bear any resemblance to typical signifiers of race, such as skin colour, ethnicity, biology or language. Still, it operates racially inasmuch it seeks to govern the potential excesses of surplus populations by assigning value to people on the basis of their capacity to overcome what is presumed to be their naturalised conditions of risk. This is not a logic that devalues the racial other *per se*, but one that revalues the racial other both for the labour it might perform and for the way it represents an important source of value for meeting the costs of adaptation. Thus, the migrant is not considered to pose a threat to capitalism but is instead said to be one that if properly managed, will sustain it. In this way, we can understand how the logic of adaptive migration converts the threat of race into a use-value for capital.

On the basis of these two examples, we can start to piece together how the logic of adaption migration, while having the appearance of being "not racist," can be understood to contribute to the adaptation of racial capitalism. It does this not only by reformulating the discourse on climate change and migration in neoliberal terms, but also, and perhaps most importantly, by rewriting race and racism. That is, the logic of adaptive migration reformulates the very terms by which race works as a technology of power. As always with race, the concept of nature is the key. Race typically defines the global division of labour in two distinctive tropes of nature, those of biology (i.e., genes or blood) or culture (i.e., timeless, prehistorical, or immutable).

But the logic of adaptive migration codifies the global division of labour racially using a different language of nature, that of environmental risk. As we saw earlier, it does this by defining environmental risk not in terms of history (e.g., the geopolitics of U.S. dollar hegemony or the history of fossil fuels) but by naturalising global climate change by defining climate change (i.e., sea-level rise, extreme weather events, drought, fire and so forth) as a set of geophysical changes to the Earth system shorn of historical substance. This new planetary technology of race differentiates people not only according to their exposure to naturalised environmental risk but according to their capacity to live with risk, their adaptive capacity. In this way, race is used to establish a break in the global division of labour between those *not* at risk of migration due to climate change, such as the northern proletariat, rights-bearing citizens, property owners and waged workers, and those who *are* at risk of migration due to climate change and, thus, who are said to pose a fundamental threat to racial capitalism but who, nevertheless, remain necessary to it. The logic of adaptive migration is a racial technology in which the latter group is governed in order to secure the interests of the former. The figure of the adaptive migrant can, thus, be understood as a racialised boundary object that holds the global division of labour together. By traversing this boundary, the adaptive migrant is imagined to undertake specific, sanctioned forms of labour in formalised labour markets (e.g., via circular and temporary migration schemes) in exchange for income that would be remitted back to informal labour markets to cover the costs of adaptation. In this way, the adaptive migrant can be understood to function as a form of risk management *for* racial capitalism. It helps to secure racial capitalism by enabling it to adapt to climate change. So, for example, global financial institutions may fear those at risk of migration due to climate change because potential migrants pose a fundament risk to capital. Those at-risk subjects may default on their debts, for example, or they may appropriate someone else's private property. Even worse, they may develop non-capitalist forms of adaptation. But rather than imposing rigid boundaries around those at risk and then subjecting them to intensive policing, the logic of adaptive migration simply recodifies those at risk of migration by revaluing them according to their adaptability. They are then monetised, accordingly, and thereby serve as sources of *both* financial speculation *and* potential sources of liquidity to enable adaptive migration. In both instances, the figure of the adaptive migrant is imagined not as a threat to capitalism but as a use-value *for* a capitalism, integral to the adaptation of capital not a harbinger of its termination. In this way, adaptive migration becomes a useful means of policing the global division of labour in the interest of maintaining racial capitalism at a critical planetary juncture in which the longevity of racial capitalism is all but guaranteed.

The logic of adaptive migration is billed as a humane corrective to more overtly racist narratives that construct migration in the terms of insecurity and humanitarian crisis. But is the logic really so innocent? Those said to be most at risk of migration due to climate change come to be understood as such because their places of residence are either no longer habitable or will soon be so. They live in conditions of searing heat, extreme drought, fire or intensive flooding, on top of generations of accumulated dispossession, slow violence, expanding levels of debt, all intensified by decades of predatory neoliberalism. It seems excessively brutal, however, to suggest that the humane option is to encourage potentially millions of vulnerable people the world over to compete in often extremely hazardous labour markets for an ever-dwindling share of wages as the price one should be expected to pay in order simply to survive on a planet they had nothing to do with destroying. But then this is precisely what neoliberalism does. It abandons vast numbers of people, who are then racialised to explain away their unliveable conditions. Moreover, if neoliberalism is a form of rationality that "disseminates the *model of the market* to all domains and activities . . . and configures human beings exhaustively as market actors,"[58] then the logic of adaptive migration certainly meets the test of being neoliberal. Not only does it displace the costs of climate change adaptation onto the world's poorest and most vulnerable, it also extends the model of the market to encompass the very conditions of survival itself. Or most crudely, it assumes that climate change adaptation is best served by creating a market for human survival in which the most willing workers are extended limited access to formalised labour markets before having to return home to unliveable circumstances. It is hard to imagine a crueller form of violence. It masquerades as a benevolent alternative to the racism inflicted by the border. Yet when we dig into the logic, it appears to enact the very worst kind of exploitation: the promise of salvation from a situation one will never be able to escape.

## RACE AND THE POLITICS OF COUNTER-TRANSFORMATION

In this chapter, I have argued that the logic of adaptive migration enacts a distinctive form of racism even while it appears to do the opposite. In making this argument, my intention is not to dismiss writers like Felli, Methmann, Oels and Bettini for whom the logic of adaptive migration is the mirror image of neoliberalism. Rather, I have sought to extend their claim by showing how the neoliberal logic of adaptive migration does not transcend race and racism but reworks both. Race is remade as a technology power that differentiates the global division of labour on the basis of environmental risk, while racism is remade as a global structure of domination that operates on and through the future, one that seeks to capture the emergent potential of the Black and brown majority at the threshold of planetary disruption.

But is adaptive migration inevitably neoliberal? Is it possible to reimagine adaptive migration as something other than the adaptation of racial capitalism? By way of a conclusion, I want to suggest that while the prevailing logic of adaptive migration found in the international political discourse on climate change and migration is neoliberal, it is not inevitably so. To assume that it is risks assigning far too much power to capital in shaping the relation between migration and climate change. Instead, drawing on insights from Nelson and Braun, we might start to rethink adaptive migration as an elemental feature of human existence which capital is simply learning to capture as a use-value for its own strategic renewal. The result is a far more radical understanding of adaptive migration, one that renders capitalism merely responsive to the conditions it creates rather than the agent of world history it is often proclaimed to be. For Braun and Nelson, capital is not a uniquely creative force in the world but is merely parasitic on the world.[59] Capital, in this sense, only ever appropriates the infinite productivity of life, including labour and migration, but it can never actually create life. Capital merely harnesses life and converts it into a fungible commodity even while it can never fully exhaust life.[60] What if adaptive migration could be understood in the same way? What if, like life itself, adaptive migration is conceptualised as infinitely productive, an example of what Nelson might call "the autonomous activity of labour power?" What kind of political possibilities may be said to reside in this reading of adaptive migration?

There is nothing inherently neoliberal about people using migration to adapt to their changing circumstances. First and foremost, what comes to be labelled "adaptive migration" within the international political discourse on climate change and migration is nothing more than the empirical phenomenon of autonomous labour mobility. Robert McLeman, an early proponent of adaptive migration, and Barry Smit, are entirely correct, for example, when they argue that the "Okies" who evacuated the Dust Bowl to settle in California in the 1930s are an historical example of those who would use migration to adapt to their changing circumstances.[61] Their point here is simply that human beings have always used migration as an adaptive strategy. I may have reservations about other moments in McLeman's empiricism. But on this fundamental point about adaptive migration, I concur with McLeman and Smit. Take, for example, the autonomous labour mobility of marronage, or the Great Migration. Both are clear examples of adaptive migration that are not neoliberal. Or consider the example of the Rohingya. Forced to migrate into Southeastern Bangladesh to escape the violence of the Myanmar military, the Rohingya have clearly used migration as an adaptation strategy that is decidedly not neoliberal. No doubt we could identify dozens of other examples in which oppressed groups of people use migration as a means of adapting to their oppressive circumstances. All are, of course, vastly different. But together they allow us to reconceptualise adaptive migration outside the framework of neoliberalism. They suggest that autonomous labour mobility is a creative, revolutionary life process that gives rise to new forms of experience, new expressions of political solidarity and possibly even new visions for the future. A full elaboration of these possibilities lies well beyond the scope of the present chapter. But if the neoliberal logic of adaptive migration is a new technology of race that is fundamentally counter-transformational, might it be possible to build a vision of climate justice that redirects the transformational potential of autonomous labour mobility and climate change to non-capitalist ends, such as the dismantling of racial capitalism?

## NOTES

1. "United Nations Discussion Panel: Insure Me: Climate Change, Human Migration, and Risk, September 24, 2009," Prevention Web, accessed 16 November 2021, https://www.preventionweb.net/events/view/11131?id=11131.

2. Koko Warner, "Human Migration and Displacement in the Context of Adaptation to Climate Change: The Cancun Adaptation Framework and Potential for Future Action," *Environment and Planning C: Government and Policy* 30, no. 6 (2012).

3. Gargi Bhattacharyya, *Rethinking Racial Capitalism: Questions of Reproduction and Survival* (London: Rowman and Littlefield, 2018); Nancy Fraser, "Roepke Lecture in Economic Geography—from Exploitation to Expropriation: Historic Geographies of Racialized Capitalism," *Economic Geography* 94, no. 1 (2018); Arun Kundnani, "What Is Racial Capitalism?," Public Lecture, Havens Wright Center for Social Justice, University of Wisconsin-Madison (15 October 2020); Arun Kundnani, "The Racial Constitution of Neoliberalism," *Race and Class* 63, no. 1 (April 2021); Cedric Robinson, *Black Marxism: The Making of the Black Radical Tradition* (Chapel Hill and London: University of North Carolina Press, 2000); Laura Pulido, "Geographies of Race and Ethnicity II: Environmental Racism, Racial Capitalism and State-Sanctioned Violence," *Progress in Human Geography* 41, no. 4 (2017); Laura Pulido, "Flint, Environmental Racism, and Racial Capitalism," *Capitalism Nature Socialism* 27, no. 3 (2016); Arun Saldanha, "A Date with Destiny: Racial Capitalism and the Beginnings of the Anthropocene," *Environment and Planning D: Society and Space* 38, no. 1 (2020).

4. Felli and Castree, "Neoliberalising"; Felli, "Managing Climate Insecurity"; Bettini, "Adaptation Strategy"; Methmann and Oels, "Fearing to Empowering."

5. Kundnani, "Racial Constitution of Neoliberalism," 64. My emphasis.

6. Robert McLeman and Barry Smit, "Migration as an Adaptation to Climate Change," *Climatic Change* 76, no. 1 (2006); Richard Black et al., "Migration as Adaptation," *Nature* 478, no. 7370 (2011); Richard Black, Neil Adger, Nigel Arnell,

Stefan Dercon, Andrew Geddes and Advid Thomas, "The Effect of Environmental Change on Migration," *Global Environmental Change* 21 (2011); Gemenne, "4 Degree Celsius+ World."

7. Giovanni Bettini and Giovanna Gioli, "Waltz with Development: Insights on the Developmentalization of Climate-Induced Migration," *Migration and Development* 5, no. 2 (2016); Methmann and Oels, "Fearing to Empowering."

8. Bettini and Gioli, "Waltz."

9. Goldberg, *The Threat of Race*.

10. Koko Warner et al., *In Search of Shelter: Mapping the Effects of Climate Change on Human Migration and Displacement* (Bonn: United Nations University, 2009).

11. Frank Laczko and Christine Aghazarm, eds., *Migration, Environment and Climate Change: Assessing the Evidence* (Geneva: International Organization for Migration, 2009).

12. Felli, "Managing Climate Insecurity." For details about the Climate Change, Environment, and Migration Alliance, see "CCEMA," United Nations University Migration Network, accessed 17 November 2021, https://migration.unu.edu/research/forced-migration/climate-change-environment-and-migration-alliance-ccema.html#outline.

13. Nash, "Negotiating Migration."

14. Felli, "Managing Climate Insecurity," 339.

15. Methmann and Oels, "Fearing to Empowering."

16. Bettini, "Adaptation Strategy"; Bettini and Gioli, "Waltz."

17. Bettini, "Adaptation Strategy," 191. The concept of "trapped populations" first appeared in the U.K. Foresight report on *Migration and Global Environmental Change*. For more on the trapped population concept, see Black and Collyer, "Populations."

18. Giovanni Bettini, "Where Next? Climate Change, Migration, and the (Bio) Politics of Adaptation," *Global Policy* 8, no. 51 (February 2017).

19. Giorgio Agamben, *State of Exception*, trans. Kevin Attell (Chicago, IL: University of Chicago Press, 2005).

20. Jeremy Walker and Melinda Cooper, "Genealogies of Resilience: From Systems Ecology to the Political Economy of Crisis Adaptation," *Security Dialogue* 42, no. 2 (April 2011).

21. Rigaud et al., *Groundswell*.

22. Mann and Wainwright, *Climate Leviathan*, 65.

23. Bettini, "Adaptation Strategy," 191.

24. This oversight, I would suggest, is largely down to a limited reading of Michel Foucault, in the case of Methmann, Oels and Bettini. Foucault is their main source of theoretical inspiration, and while he occupies an ambivalent place in contemporary race analysis, Foucault nevertheless had much to say about race and biopower. In contrast, I attribute Felli's oversight of race and racism down to his reading of Marxist political economy, which prioritises class over race as the foundational concept of difference animating capitalist production. In pointing out this shortcoming, my point is not to condemn these authors. It does, however, highlight a fundamental debate in

contemporary political theory concerning the relative significance different authors place on race and class as key organising categories under capitalism. Foucault's treatment of race can be found in his lectures at the Collège de France published as *"Society Must Be Defended": Lectures at the Collège de France, 1975–1976* (New York: Picador, 2003) and *Abnormal: Lectures at the Collège de France, 1974–1975* (New York: Picador, 2003). Race also figures in the *History of Sexuality: An Introduction, Volume 1* (Random House, 1978; New York: Vintage Books, 1990). See also Stoler, *Race*; Goldberg, *Racial State*; Goldberg, *Racist Culture.*

25. Bruce Braun, "New Materialisms and Neoliberal Natures," *Antipode* 47, no. 1 (January 2015); Sara Nelson, "Resilience and the Neoliberal Counter-Revolution: From Ecologies of Control to the Production of the Common," *Resilience* 2, no. 1 (January 2014); Sara Nelson, "Beyond the Limits to Growth: Ecology and the Neoliberal Counterrevolution," *Antipode* 47, no. 2 (2015).

26. Robinson, *Black Marxism.*

27. Pulido, "Geographies of Race"; Pulido, "Flint."

28. Ahuja, *Planetary Specters.*

29. Kundnani, "Racial Constitution of Neoliberalism," 52.

30. Ibid., 52. My emphasis.

31. Kundnani, "Racial Constitution of Neoliberalism." For more on Hayek's contribution to neoliberal philosophy, see, for example, Walker and Cooper, "Genealogies of Resilience"; Michael Watts, "On Confluences and Divergences," *Dialogues in Human Geography* 1, no. 1 (2011); Melinda Cooper, "Complexity Theory after the Financial Crisis: The Death of Neoliberalism or the Triumph of Hayek?," *Journal of Cultural Economy* 4, no. 4 (2011).

32. Kundnani, "Racial Constitution of Neoliberalism," 60.

33. Ibid., 60.

34. Étienne Balibar, "Is There a Neo-Racism?," in *Race, Nation, Class: Ambiguous Identities*, ed. Étienne Balibar and Immanuel Wallerstein (London: Verso, 1991).

35. Kundnani, "Racial Constitution of Neoliberalism," 63. For his part, Goldberg argues that the implementation of neoliberalism, at least in America, was partly motivated to scale back the size of the state because it provided a minimal level of welfare for poor Americans, many of whom were African American, and also because, under affirmative action hiring, the state had become the single largest employer of people of African descent in the United States. Thus, neoliberalism, for Goldberg, can be interpreted as a response to what many white Americans at the time felt was "the impending impotence of whiteness" signified by the welfare state. See Goldberg, *Threat of Race*, 337. On racial neoliberalism, see also, David Roberts and Minelle Mahtani, "Neoliberalizing Race, Racing Neoliberalism: Placing 'Race' in Neoliberal Discourses," *Antipode* 42, no. 2 (2010).

36. Kundnani, "Racial Constitution of Neoliberalism," 62.

37. Ibid., 64.

38. Ibid.

39. Ibid., 53.

40. Ibid., 65.

41. Ibid., 64–65.

42. Ibid., 65.

43. Kelley et al., "Fertile Crescent."

44. Ahuja, *Planetary Specters*.

45. At the time of writing, the United Nations Framework Convention on Climate Change Conference of the Parties has been meeting for twenty-six years to resolve the climate crisis. It was founded in a moment of tremendous optimism in the early 1990s but has consistently failed to meet its ambitions. Meanwhile, the crisis only seems to be getting worse.

46. Ahuja, *Planetary Specters*.

47. Robert Vitalis's history of Aramco explains in depth how the labour practices used by the company in Saudi Arabia were modelled on Jim Crow segregation in America. See Robert Vitalis, *America's Kingdom: Mythmaking on the Saudi Oil Frontier* (Stanford, CA: Stanford University Press, 2008).

48. Kasia Paprocki, "All that is solid melts into the bay: Anticipatory ruination and climate change adaptation," *Antipode* 51, no. 1 (2019): 295–315.

49. White, *Climate Change and Migration.*

50. U.K. Foresight, *Migration*, 120. My emphasis.

51. Asian Development Bank, *Addressing Climate Change and Migration in Asia.*

52. Government of the People's Republic of Bangladesh, *Bangladesh Climate Change Strategy and Action Plan* (Dhaka: Ministry of Environment and Forests, 2009), 17, as quoted in Ahuja, *Planetary Specters*, 113.

53. Asian Development Bank, *Addressing Climate Change and Migration in Asia.*

54. For an overview of catastrophe bonds and security, see Luis Lobo-Guerrero, *Insuring Security: Biopolitics, Security and Risk* (London: Routledge, 2011); Leigh Johnson, "Catastrophe Bonds and Financial Risk: Securing Capital and Rule through Contingency," *Geoforum* 45 (March 2013); Leigh Johnson, "Geographies of Securitized Catastrophe Risk and the Implications of Climate Change," *Economic Geography* 90, no. 2 (2014). On weather derivatives, see Jane Pollard et al., "Firm Finances, Weather Derivatives, and Geography," *Geoforum* 39, no. 2 (March 2008); Sam Randalls, "Weather Profits: Weather Derivatives and the Commercialization of Meteorology," *Social Studies of Science* 40, no. 5 (2010).

55. Asian Development Bank, *Addressing Climate Change and Migration in Asia*, 62.

56. Space precludes a full discussion of what I mean by climate justice in this context. For simplicity, I am simply referring to broad-based social movements that seek to ensure the resolution to the climate crisis is accountable to and, thus, reverses the effects of global histories of racial violence.

57. Kundnani, "Racial Constitution of Neoliberalism," 53.

58. Wendy Brown, *Undoing the Demos: Neoliberalism's Stealth Revolution* (Brooklyn, NY: Zone Books, 2015), 31. Emphasis in original.

59. Toni Morrison identifies the "nature of white freedom" as "*parasitic*" in *Playing in the Dark* (my emphasis). See Toni Morrison, *Playing in the Dark: Whiteness and the Literary Imagination* (London: Picador, 1993), 52. Alison Howell and Melanie Richter-Monpetit develop this idea as does Frank Wilderson III. See Howell and Richter-Monpetit, "Racism in Foucauldian Security Studies: Biopolitics, Liberal

War, and the Whitewashing of Colonial and Racial Violence," *International Political Sociology* 13, no. 1 (2019); Wilderson, *Afropessimism*, 94.

60. Cooper, *Life as Surplus*; Elizabeth Johnson and Jesse Goldstein, "Biomimetic Futures: Life, Death, and the Enclosure of a More-than-Human Intellect," *Annals of the Association of American Geographers* 105, no. 2 (2015).

61. McLeman and Smit, "Adaptation to Climate Change."

# Conclusion

In the days leading up to COP26 in Glasgow, U.K. prime minister Boris Johnson likened climate change to the fall of Rome.[1] His point was that if climate change goes unaddressed, the barbarians will soon arrive on the frontiers of Europe and lead Western civilisation into a new dark age.[2] We might be tempted to dismiss Johnson's remark as that of a beleaguered populist with a knack for classicist turns of phrase. Yet his comment is one, I think, we should take very seriously. Not, however, because it contains some epochal truth. But because it alerts us, once again, to the constitutive presence of the other within the climate change imaginary. It reminds us that for all the critique we might direct at the figure of the climate migrant/refugee, for all the effort spent debunking the mythology of "climate change and migration,"[3] the figure remains an irresistible leitmotif within the climate change political thought. It may appear in many guises. The end of Western civilisation, for example, or a crisis of human rights. And it may well provoke a range of reactions: higher border walls, longer fences, more intensive policing, legal protections, financialisation and new mechanisms of labour control. But even while these conceptions of the other of climate change differ in important respects, all agree on one thing: global climate change stands to displace millions of people around the world. What makes Johnson's trope of civilisational decline so significant, I would suggest, is that it reformulates this generalised anxiety about the other of climate change into a public narrative about shared political community. It seeks to redefine the publicness of climate change—the idea that climate change is something *everyone* has in common—in terms of an uncanny other that marks the outer contours of the climate change imaginary. On one side of this imaginary stands a political community that for all its contradictions and internal squabbles finds agreement around notions of civilisation, humanism, right, stability and historical agency. While on the other, we find a chimera, defined, paradoxically, by its indefinability. How we come to terms with this structure of difference, this "architecture of enmity," remains one of the most pressing tasks of critical thought for our times.[4]

*The Other of Climate Change* has sought to address this challenge by bringing urgent, critical attention to the structural outside of this shared political imaginary. My argument here has been twofold. First, I have argued that the figure of the climate migrant/refugee stands for the revival of Western humanism at the very moment humanism is now having to confront a global environmental crisis whose origins are humanism itself. The human is coming unglued. The planet it sought to modernise, domesticate and cajole into submission, not least through a history of imperial and colonial conquest, is now speaking back with a vengeance.[5] For the first time in its history, the human is now having to navigate a seismic tension between itself as a geophysical *force* on the planet and the *forcings* of the planet itself.[6] But that we are unable to decipher between the two is what makes our planetary conjuncture so bewildering. Where does human force end and planetary forcing begin? My contention is that the condition of possibility for the figure of the climate migrant/refugee is precisely this disorienting convergence of force and forcing. Without a readymade map for finding our way through this unique planetary condition, we instead seem intent to rely on a tried-and-tested strategy for resolving our own internal crises. The human—us—invents an other—in this case the figure of the climate migrant/refugee. We offer up the other of climate change as the boundary marker between force and forcing. We ask this unassuming other to resolve the crisis, just as we assign it the useful purpose of dissimulating the origins of the crisis itself. We project onto this other all our troubled fantasies about the crisis, just as we seek to extract from it an ethics, moral logic and politics of response. The figure is not a real presence in the world, but something humanism has had to invent in order to come to terms with a crisis it alone has created. It is, I would suggest, the very means by which humanism is reconsolidated in the face of planetary upheaval. Thus, my first argument is that the figure is not just the constitutive outside of a shared political community called "western humanism." It has also become a necessary supplement to humanism's strategic adaptation to climate change.

Equally important, though, I have argued that the political imaginary whose founding exclusion is the other of climate change is racial or, more precisely, anti-Black. That is, the political community the other serves to limn is not simply that of humanism but of *white* humanism. White humanism is the position of unmarked universality from which the other of climate change is constructed. It is from this position that the touchpoint between the empirical fact of climate change, as a set of geophysical impacts, and the historical life of migration, is imagined. If there is a knowledge of this touchpoint, whether it takes the form of scenarios, science fiction, speculative ethics or aesthetics, this knowledge is only ever possible because of a prior white humanism whose operative logic is the production of Blackness. The discourse on climate change and migration merely repeats white humanism's anti-Black operative logic. One might trace the repetition of this logic through specific national registers. But the specific genre of white humanism that concerns me in this book is that which takes shape in relation to the international (chapter 3). This reflects the fact that the political community of climate change science and policy is itself global and international as well as the fact that international institutions play an important part in shaping the discourse on migration and climate change. And it is *white* because it constructs the other in accordance with a set of racial tropes that position the other outside the category of the human (chapter 2). In this way, those who are said to be on the verge of being displaced by climate change, most notably the poor who make up the brown and Black global precariat, are made to occupy the foundational nothingness that has *always* been the marker of Blackness.[7] And this is significant because Blackness has itself always served as a source of meaning and replenishment for white humanism.[8] From this perspective, the adaptation of white humanism is only possible in relation to and because of its Black other. At issue here, though, is not whether the other is perceived as either "good" or "bad." Nor is the remedy to this logic to be found in a vocabulary that will define the figure of the climate migrant/refugee in non-racist terms once and for all. There is no other of climate change that can be grasped outside the framework of racism. What matters above all is that the other of climate change, an historical articulation of Blackness, is foundational to the political community of white humanism and, thus foundational, to its adaptation to climate change. Consequently, my second argument is that, within the discourse on climate change and migration, we find not simply the adaptation of white humanism but the adaptation of race and racism.

International climate change political thought is, therefore, not immune from considerations of race and racism. After the critique of the figure of the climate migrant/refugee, the science and policy discourse on climate change can no longer be thought of as race-neutral. My hope is that *The Other of Climate Change* makes this point abundantly clear. Readers should understand, however, that, by taking this position, my purpose in writing this book has not been to level some grand indictment against those seeking to mitigate climate change. Much less have I sought to incriminate those seeking justice for some of the poorest and most vulnerable to climate change. Instead, my purpose in writing this book has been to draw much needed attention to the fact that race and racism are intimately bound up with what it means to mitigate and adapt to climate change. Solving the crisis of climate change on behalf of all of humanity, however well intentioned, cannot be decoupled from the historical particularity of white humanism. For while the climate change imaginary is ostensibly universal, the figure of the climate migrant/

refugee reveals it to be a white imaginary by another name—just another epistemological site in the repetition of white humanism. For many readers, the point I am trying to stress here is obvious: interrogating the other of climate change brings into focus how the universality of the climate change imaginary is an historical manifestation of whiteness. While for other readers, the whiteness of climate change will be much harder to see and, therefore, much harder to accept. Whether and how the international climate change policy and science community comes to terms with this observation remains to be seen. But come to terms it must.

Among the many risks in doing so, however, is that the international community may follow the path of least resistance, falling back on old conceptions of race and racism that underpinned the post-war compromise. This would be unfortunate. It might, for example, adopt the argument that race is nothing but a "social construction." Or it might offer reassuring pieties that human difference ought to be celebrated or that human flourishing depends on the rich tapestry of humanistic diversity. Such were the narratives the international community would adopt in the early 1950s to move past the racial atrocities of the twentieth century.[9] But even as these narratives would officially discredit racial science, they would simply reformulate race by substituting culture for biology in what would become the primary means for naturalising difference in the post-war era, giving rise to what Étienne Balibar would call neo-racism.[10] Similarly, the international community might seek to rectify the whiteness of climate change by narrowly associating racism with the "far-right," border security and militarism, even as it continues to promote the logic of adaptive migration (chapter 6). But all this would do is overlook the fact that racism is a system of signification whose primary function has only ever been to differentiate Europeans from non-Europeans. It is this historical fact the international community must now confront.

The stakes here are, of course, massive. Coming to terms with race and racism in the context of our present crisis is no trivial matter. It is, therefore, all the more urgent that those working to bring about a "global climate justice revolution" think carefully about how the figure of the climate migrant/refugee shapes the meaning of climate change politics and practice.[11] Not only is the figure central to the adaptation of humanism, but it is also, I would argue, central to what Mann and Wainwright call the "adaptation of the political."[12] The political is, for them, the arena of capitalist political economy "in which the relations between the dominant and dominated are worked out."[13] For my part, the international political discourse on climate change and migration is one of the primary sites, maybe even *the* site, where we can locate this "working out." At stake here is nothing less than an effort to redefine the political contours of white humanism. But this "working out" is only partly about defining who falls within the category of the climate migrant/refugee, only partly about deciding who will survive climate change and on what terms. Even more important, the discourse on climate change and migration establishes who has the power to draw these distinctions in the first place. This is a power that establishes the authority to make the other stand as the boundary marker between human and inhuman worlds. Or most straightforwardly, it is a power to redraw the boundaries of race and racism for the era of climate change. In an era of intensifying anti-immigration rhetoric, it would be easy to conclude that nation states and their border agencies have come to play a leading role in defining the terms of the discourse. What I have tried to show, however, is that much of this discourse on climate change and migration takes shape within the scope of international institutions. Developing analytical approaches that can hold these institutions to account, thus, continues to be a vital task for any political theory of climate change.

My hope is that the concept of racial futurism offers something to this task. Racial futurism is a unique form of racial rule. It characterises the figure of the climate migrant/refugee by using a set of distinctively racial tropes (chapter 2). Its primary site of articulation is the white international (chapter 3), and it follows a racial logic that mimics the post-racial by conceptualising racism in the narrowest possible terms. It does this by defining racism as something that by virtue of having occurred in the distant past no longer has relevance to climate change politics in the present. In this way, it conforms to the paradoxical racist logic of "not racism" (chapter 4), an expression of racism without racism for climate change.

But what really sets racial futurism apart from other modalities of racial rule (e.g., racial historicism and racial naturalism) is that it conceptualises its racial other as that which may materialise in the future. The result is a modality of rule whose other is permanently deferred and whose presence is defined in terms of its virtuality. This is a power that lies in fostering a racial orientation to climate change (chapter 5), while its logic is one that seeks to ward off this future by redoubling the effort to control global surplus labour in the present. Racial futurism will almost certainly amount to ever-tightening border restrictions. Yet racial futurism is irreducible to the logic of the border. Just as significant, it is a speculative logic that seeks to capture the *risk* of migration by converting those most at risk of displacement into a use- value for capital (chapter 6).

The conception of racial futurism that I elaborate in *The Other of Climate Change* is akin to what Brian Massumi has called ontopower. Ontopower is concerned with the "regulation of effects, rather than of causes," but the regulation of effects *before* such effects become fully actualised.[14] Racial futurism operates in much the same way. As an anticipatory form of racism, it codifies and regulates the racial other *before* it arises, prior to its actual materialisation as such. We can think of racial futurism as a form of racism that both produces and acts on the *virtuality* of the racial other rather than on actual events of migration. Its other is not a material thing in the world—a climate migrant, for example—only the felt sense of the other's presence. Racial futurism is a power that works by endlessly repeating the future event of migration without resolution, by producing and sustaining a sense of the other's irresolvable presence. This is why the scenarios discussed in chapter 5 are so consequential. Scenarios don't define the figure of the climate migrant/refugee in any material sense. They do, however, confirm the figure's *objective plausibility*. They help to generate a sense that climate-induced migration could become a problem but without making any empirical commitments to the actuality of the thing itself. It is precisely this sense-making that is racial futurism's "operative logic."

An operative logic "combines an ontology with an epistemology in such a way as to endow it with powers of self-causation."[15] It is, as Massumi would put it, the "process of becoming formative of its own species being."[16] "What does an operative logic want," Massumi asks? His answer: "Its own continuance."[17] Racial futurism is white humanism's operative logic. It remakes humanism on the threshold of climate change. Most fundamentally, racial futurism specifies the terms by which race and racism are adapting to climate change. But racial futurism is also, paradoxically, a plea for survival. Not, however, a plea for those who would suffer the indignity of being overwhelmed by climate change, but a plea for the human itself whose unbounded fantasy for the other lies at the heart of the climate crisis. Racial futurism is a differentiating power that works as humanism's last defence against climate change. This is the order of governance humanism seeks to preserve for itself and which is at stake in the adaptation of the political.

The concept of *the other* is of fundamental importance for a critical reading of contemporary climate change. For Étienne Balibar, the concept of the other is a civilisational marker.[18] To know the other is at once a form of power/knowledge *and* civilisational pedagogy. We come to know the other, and through knowing the other we come to a clearer sense of "who we are." This was, of course, the lesson of Edward Said. But it applies just equally to the climate crisis; as we come to know the climate change other, so we come to a fuller sense of who we are in the face of a full-on planetary crisis. So for Balibar, "the Other is not really or purely exterior. It is also interior, constitutive of oneself. Without the other there would be no possibility of civilizing oneself."[19] The other of climate change may well signify the desire to subdue the other "out there." But the lesson I think we should take from Balibar is that the other of climate change also signifies the desire to subdue the other that resides within each of us. "The construction of the Other is the construction of an *alienated Self*, where all the properties attributed to the

Other are inversions and distortions of those vindicated for oneself."[20] Racial futurism may well signify a fantasy of racial domination. But to me a far more satisfying reading is that the other of climate change announces white humanism's desire to quell its own unruly impulses, a desire to ward off its own becoming-Black. To paraphrase James Baldwin, we might, therefore, ask why we need the figure of the climate migrant/refugee in the first place.[21] What is it about humanism at the threshold of climate change that can explain the desire to rule over this other? Any unthinking of race and racism in relation to climate change should begin with this basic question.

## NOTES

1. Tim Ross, "Boris Johnson Likens Climate Change to 'the Decline and Fall of the Roman Empire,'" *The New Statesman*, 29 October 2021.
2. For an excellent treatment of the barbarian trope, see Bettini, "Barbarians."
3. Ingrid Boas et al., "Climate Migration Myths," *Nature Climate Change* 9, no. 12 (2019).
4. Michael Shapiro uses the term "architecture of enmity" to describe how differences are drawn as a precondition for war. See Michael Shapiro, *Violent Cartographies: Mapping Cultures of War* (Minneapolis: University of Minnesota Press, 1997). Also, Louise Amoore, "Algorithmic War: Everyday Geographies of the War on Terror," *Antipode* 41, no. 1 (2009); and Gregory, *The Colonial Present*.
5. Isabelle Stengers, *In Catastrophic Times: Resisting the Coming Barbarism* (London: Open Humanities Press, 2015).
6. Baucom, *History 4° Celsius*, 8.
7. Warren, *Ontological Terror.*
8. Wilderson, *Afropessimism*.
9. Étienne Balibar, "Difference, Otherness, Exclusion," *Parallax* 11, no. 1 (2005).
10. Balibar, "Neo-Racism."
11. Mann and Wainwright, *Climate Leviathan*, 18.
12. Ibid., 79–98.
13. Ibid., 19.
14. Massumi, *Ontopower*, 22.
15. Ibid., 200.
16. Ibid., 201.
17. Ibid.
18. Balibar, "Difference, Otherness, Exclusion."
19. Ibid., 30.
20. Ibid.
21. *I Am Not Your Negro*, directed by Raoul Peck (Velvet Film & Magnolia Pictures, 2016) 1:33:00, https://www.imdb.com/title/tt5804038/.

# Bibliography

Agamben, Giorgio. *Homo Sacer: Sovereign Power and Bare Life.* Translated by Daniel Heller-Roazen. Stanford, CA: Stanford University Press, 1998.

———. *State of Exception.* Translated by Kevin Attell. Chicago, IL: University of Chicago Press, 2005.

Ahmed, Sara. "Declarations of Whiteness: The Non-Performativity of Anti-Racism." *Borderlands* 3, no. 2 (2004).

———. "A Phenomenology of Whiteness." *Feminist Theory* 8, no. 2 (2007): 149–68.

———. *Strange Encounters: Embodied Others in Post-Coloniality.* London: Routledge, 2000.

Ahuja, Neel. *Planetary Specters: Race, Migration, and Climate Change in the Twenty-First Century.* Chapel Hill: University of North Carolina Press, 2021.

Alcoff, Linda *The Future of Whiteness.* London: Polity Press, 2015.

Amin, Ash. "The Remainders of Race." *Theory, Culture, and Society* 27, no. 1 (March 2010): 1–23.

Amoore, Louise. "Algorithmic War: Everyday Geographies of the War on Terror." *Antipode* 41, no. 1 (2009): 49–69.

———. "Security and the Incalculable." *Security Dialogue* 45, no. 5 (October 2014): 423–39.

Anderson, Ben. "Affect and Biopower: Towards a Politics of Life." *Transactions of the Institute of British Geographers* 37, no. 1 (2012): 28–43.

———. "Preemption, Precaution, Preparedness: Anticipatory Action and Future Geographies." *Progress in Human Geography* 34, no. 6 (2010): 777–98.

Anderson, Ben and Rachel Gordon. "Government and (Non)Event: The Promise of Control." *Social and Cultural Geography* 18, no. 2 (2017): 158–77.

Anderson, Kay. *Race and the Crisis of Humanism.* London: Routledge, 2005.

Anievas, Alexander, Nivi Manachanda and Robbie Shilliam, eds. *Race and Racism in International Relations.* London: Routledge, 2015.

Asian Development Bank. *Addressing Climate Change and Migration in Asia and the Pacific.* Manila: Asian Development Bank, 2012.

Atapattu, Sumudu. "Climate Change: Disappearing States, Migration, and Challenges for International Law." *Washington Journal of Environmental Law & Policy* 4, no. 1 (2014): 1–36.

Baldwin, Andrew. "Orientalising Environmental Citizenship: Climate Change, Migration and the Potentiality of Race." *Citizenship Studies* 16, no. 5–6 (2012), 625–640.

———. "Premediation and White Affect: Climate Change and Migration in Critical Perspective." *Transactions of the Institute of British Geographers* 41, no. 1 (2016): 78–90.

———. "Racialisation and the Figure of the Climate Change Migrant." *Environment and Planning A* 45, no. 6 (2013): 1474–90.

———. "Resilience and Race, or Climate Change and the Uninsurable Migrant: Towards an Anthroporacial Reading of Race." *Resilience: International Policies, Practices, and Discourses* 5, no. 2 (2017): 129–42.

Balibar, Étienne. "Difference, Otherness, Exclusion." *Parallax* 11, no.1 (2005): 19–34.

———. "Is There Is a Neo-Racism?" In *Race, Nation, Class: Ambiguous Identities*, edited by Étienne Balibar and Immanuel Wallerstein, 17–36. London: Verso, 1991.

Bannerji, Himani. *The Dark Side of the Nation: Essays on Multiculturalism, Nationalism and Gender*. Toronto: Canadian Scholars Press, 2000.

Baucom, Ian. *History 4° Celsius: Search for a Method in the Age of the Anthropocene*. Durham, NC, and London: Duke University Press, 2020.

Barnett, Jon. "Security and Climate Change." *Global Environmental Change* 13, no. 1 (2003): 7–17.

Bennett, Jane. *VibrantMatter: A Political Ecology of Things.* Durham, NC, and London: Duke University Press, 2010.

Bettini, Giovanni. "Climate Barbarians at the Gate? A Critique of Apocalyptic Narratives on 'Climate Refugees.'" *Geoforum* 45 (2013): 63–72.

———. "Climate Migration as an Adaptation Strategy: De-Securitizing Climate-Induced Migration or Making the Unruly Governable?" *Critical Studies on Security* 2, no. 2 (2014): 180–95.

———. "(In)Convenient Convergences: 'Climate Refugees,' Apocalyptic Discourses and the Depoliticization of Climate-Induced Migration." In *Deconstructing the Greenhouse: Interpretive Approaches to Global Climate Governance*, edited by Chris Methmann, Delf Rothe and Benjamin Stephan, 122–36. Abingdon: Routledge, 2013.

———. "Where Next? Climate Change, Migration, and the (Bio)Politics of Adaptation." *Global Policy* 8 no. 51 (February 2017): 33–39.

Bettini, Giovanni and Giovanna Gioli. "Waltz with Development: Insights on the Developmentalization of Climate-Induced Migration." *Migration and Development* 5, no. 2 (2016): 171–89.

Bhambra, Gurminder. "Brexit, Trump, and 'Methodological Whiteness': On the Misrecognition of Race and Class." *The British Journal of Sociology* 68, no. S1 (2017), S214–S232.

Bhattacharyya, Gargi. *Rethinking Racial Capitalism: Questions of Reproduction and Survival*. London: Rowman and Littlefield, 2018.

Biermann, Frank and Ingrid Boas. "Preparing for a Warmer World: Towards a Global Governance System to Protect Climate Refugees." *Global Environmental Politics* 10, no. 1 (2010): 60–88.

Black, Richard. "Environmental Refugees: Myth or Reality?" In *New Issues in Refugee Research: Working Paper 34.* Geneva: United Nations High Commissioner for Refugees, 2001.

Black, Richard, Dominic Kniveton and Kerstin Schmidt-Verkerk. "Migration and Climate Change: Towards an Integrated Assessment of Sensitivity." *Environment and Planning A* 43, no. 2 (2011): 431–50.

Black, Richard and Michael Collyer. "Populations 'Trapped' at Times of Crisis." *Forced Migration Review* 45 (2014): 52–56.

Black, Richard, Neil Adger, Nigel Arnell, Stefan Dercon, Andrew Geddes and David Thomas. "The Effect of Environmental Change on Migration." *Global Environmental Change* 21, no. S1 (December 2011): S3–S11.

Black, Richard, Stephen Bennett, Sandy Thomas and John Beddington. "Migration as Adaptation." *Nature* 478, no. 7370 (2011): 447–49.

Boano, Camillo, Roger Zetter and Tim Morris. *Environmentally Displaced People: Understanding the Linkages between Environmental Change, Livelihoods and Forced Migration.* Refugee Studies Centre, Oxford University, Department of International Development, 2008.

Boas, Ingrid. *Climate Migration and Security: Securitisation as a Strategy in Climate Change Politics*. New York and London: Routledge, 2015.

Boas, Ingrid, Carol Farbotko, Helen Adams, Harald Sterly, Simon Bush, Kees van der Geest, Hanne Wiegel, Hasan Ashraf, Andrew Baldwin, Giovanni Bettini, Suzy Blondin, Mirjam de Bruijn, David Durand-Delacre, Christiane Fröhlich, Giovanna Gioli, Lucia Guaita, Elodie Hut, Francis X. Jarawura, Machiel Lamers, Samuel Lietaer, Sarah L. Nash, Etienne Piguet, Delf Rothe, Patrick Sakdapolrak, Lothar Smith, Basundhara Tripathy Furlong, Ethemcan Turhan, Jeroen Warner, Caroline Zickgraf, Richard Black and Mike Hulme. "Climate Migration Myths." *Nature Climate Change* 9, no. 12 (2019): 901–03.

Bogardi, Janos and Koko Warner. "Here Comes the Flood." *Nature Climate Change* 138, no. 11 (2008): 9–11.

Bonilla-Silva, Eduardo. *Racism without Racists: Color-Blind Racism and the Persistence of Racial Inequality in America.* Lanham, MD: Rowman and Littlefield, 2018.

Bonnett, Alastair. "Geography, 'Race,' and Whiteness: Invisible Traditions and Current Challenges." *Area* 29, no. 3 (1997): 193–99.

Braun, Bruce. *The Intemperate Rainforest: Nature, Culture, and Power on Canada's West Coast*. Minneapolis: University of Minnesota Press, 2002.

———. "New Materialisms and Neoliberal Natures." *Antipode* 47, no. 1 (January 2015): 1–14.

Brown, Lester. "The Rising Tide of Environmental Refugees." Washington, DC: InterPress Service News Agency 22 (October 2009).

———. *World on the Edge: How to Prevent Environmental and Economic Collapse.* New York and London: W.W. Norton & Co., 2011.

Brown, Oli. "The Numbers Game." *Forced Migration Review* 31 (October 2008): 8–9.

Brown, Wendy. "Climate Change, Democracy and Crises of Humanism." In *Life Adrift: Climate Change, Migration, Critique*, edited by Andrew Baldwin and Giovanni Bettini, 25–40. London: Rowman and Littlefield, 2017.

———. *Regulating Aversion: Tolerance in the Age of Identity and Empire*. Princeton, NJ: Princeton University Press, 2006.

———. *Undoing the Demos: Neoliberalism's Stealth Revolution*. Brooklyn, NY: Zone Books, 2015.

———. *Walled States, Waning Sovereignty*. Brooklyn, NY: Zone Books. 2010.

Buchanan, Kelly. "New Zealand: 'Climate Change Refugee' Case Overview." Edited by Law Library of Congress, Global Legal Research Center, 2015.

Burkett, Maxine. "The Nation Ex-Situ: On Climate Change, Deterritorialized Nationhood and the Post-Climate Era." *Climate Law* 2 (2011): 345–74.

Butler, Judith. *Bodies That Matter: On the Discursive Limits of "Sex."* New York: Routledge, 1993.

———. *Gender Trouble: Feminism and the Subversion of Identity*. New York: Routledge, 1990.

Campbell, Kurt M., Jay Gulledge, John R. McNeill, John Podesta, Peter Ogden, Leon Fuerth, R. James Woolsey, Alexander T. Lennon, Julianne Smith and Richard Weitz. *The Age of Consequences: The Foreign Policy and National Security Implications of Global Climate Change*. Washington, DC: Center for Strategic and International Studies and Center for a New American Security, 2007.

Castree, Noel. *Making Sense of Nature: Representation, Politics and Democracy*. Abingdon: Routledge, 2014.

———. *Nature*. London: Routledge, 2005.

Chakrabarty, Dipesh. "Postcolonial Studies and the Challenge of Climate Change." *New Literary Review* 43, no. 1 (2012): 1–18.

Chaturvedi, Sanjay and Timothy Doyle. *Climate Terror: A Critical Geopolitics of Climate Change*. Basingstoke: Palgrave Macmillan, 2015.

Chowdhry, Geeta and Sheila Nair. "Introduction: Power in a Postcolonial World: Race, Gender, and Class in International Relations." In *Power, Postcolonial and International Relations: Reading Race, Gender and Class*, edited by Geeta Chowdhry and Sheila Nair, 1–32. London: Routledge, 2004.

Chowdhry, Geeta and Sheila Nair, eds. *Power, Postcolonialism and International Relations: Reading Race, Gender and Class*. London: Routledge, 2004.

Christian Aid. *Human Tide: The Real Migration Crisis*. London: Christian Aid, 2007.

Clement, Viviane, Kanta Kumari Rigaud, Alex de Sherbinin, Bryan Jones, Susana Adamo, Jacob Schewe, Nian Sadiq and Elham Shabahat. *Groundswell Part 2: Acting on Internal Climate Migration*. Washington, DC: The World Bank, 2021.

CNA. *National Security and the Threat of Climate Change*. Arlington: CNA Corporation, 2007.

Coates, Ta-Nehisi. *Between the World and Me*. Melbourne: Text Publishing, 2015.

Collectif Argos. *Climate Refugees*. Singapore: Tien Wah Press, 2010.

Collier, Stephen. "Topologies of Power: Foucault's Analysis of Political Government beyond 'Governmentality.'" *Theory, Culture, and Society* 26, no. 6 (2009): 78–108.

Cooper, Melinda. "Complexity Theory after the Financial Crisis: The Death of Neoliberalism or the Triumph of Hayek?" *Journal of Cultural Economy* 4, no. 4 (2011): 371–85.

———. *Family Values: Between Neoliberalism and the New Social Conservativism*. Brooklyn, NY: Zone Books, 2017.

———. "Insecure Times, Tough Decisions: The Nomos of Neoliberalism." *Alternatives* 29 (2004): 515–33.

———. *Life as Surplus: Biotechnology and Capitalism in the Neoliberal Era*. Seattle: University of Washington Press, 2007.

———. "Turbulent Worlds: Financial Markets and Environmental Crisis." *Theory, Culture and Society* 27, no. 2–3 (2010): 167–90.

Coulthard, Glen. *Red Skin, White Masks: Rejecting the Colonial Politics of Recognition*. Minneapolis: University of Minnesota Press, 2014.

Dalby, Simon. *Security and Environmental Change*. Cambridge: Polity, 2009.

De Goede, Marieke. "Beyond Risk: Premediation and the Post 9/11 Security Imagination." *Security Dialogue* 39, no. 2–3 (2008): 155–76.

De Goede, Marieke and Sam Randalls. "Precaution, Preemption: Arts and Technologies of the Actionable Future." *Environment and Planning D: Society and Space* 27, no. 5 (January 2009): 859–78.

De Goede, Marieke, Stephanie Simon and Marijn Hoijtink. "Performing Preemption." *Security Dialogue* 45, no. 5 (2014): 411–22.

Deleuze, Gilles. *Difference and Repetition*. New York: Columbia University Press, 1994.

Dibley, Ben and Brett Neilson. "Climate Crisis and the Actuarial Imaginary: 'The War on Global Warming.'" *New Formations* 69 (2010): 144–59.

Dillon, Michael and Julian Reid. *The Liberal Way of War: Killing to Make Life Live*. London: Routledge, 2009.

Docherty, Bonnie and Tyler Giannini. "Confronting a Rising Tide: A Proposal for a Convention on Climate Change Refugees." *Harvard Environmental Law Review* 33, no. 2 (2009): 349–403.

Drayton, Richard. "The Road to the Tricontinental: Anti-Imperialism and the Politics of the Twentieth Century." Conference Paper. *Legacies of the Tricontinental: Imperialism, Resistance, Law*. 22–24 September 2016. Coimbra: Centre for Social Studies.

Du Bois, W. E. B. *The Souls of Black Folk*. London: Archibald Constable & Co., 1905.

Duffield, Mark. *Development, Security, and Unending War: Governing the World of Peoples*. London: Polity Press, 2007.

———. "Racism, Migration and Development: The Foundations of Planetary Order." *Progress in Development Studies* 6, no. 1 (January 2006): 68–79.

Dun, Olivia and François Gemenne. "Defining 'Environmental Migration.'" *Forced Migration Review* 31 (October 2008): 10–11.

Dwyer, Owen and John Paul Jones III. "White Socio-Spatial Epistemology." *Social and Cultural Geography* 1, no. 2 (2000): 209–22.

Dyer, Richard. *White*. London and New York: Routledge, 1997.

Eddo-Lodge, Reni. *Why I Am No Longer Talking to White People about Race.* London: Bloomsbury, 2018.

Elden, Stuart. *Terror and Territory: The Spatial Extent of Sovereignty*. Minneapolis: University of Minnesota Press, 2009.

El-Hinnawi, Essam. *Environmental Refugees*. Nairobi: United Nations Environment Programme, 1985.

Faber, Daniel and Christina Schlegel. "Give Me Shelter from the Storm: Framing the Climate Refugee Crisis in the Context of Neoliberal Capitalism." *Capitalism Nature Socialism* 28, no. 3 (2017): 1–17.

Fanon, Frantz. *Black Skin, White Masks* (Paris: Editions de Seuil, 1952). Translated by Charles L. Markmann. London: Pluto Press, 2017.

———. *The Wretched of the Earth.* (Paris: François Maspéro, 1961). Translated by Constance Farrington. London: Penguin Books, 2001.

Farbotko, Carol. "'The Global Warming Clock Is Ticking So See These Places While You Can': Voyeuristic Tourism and Model Environmental Citizens on Tuvalu's Disappearing Islands." *Singapore Journal of Tropical Geography* 31 (2010): 224–38.

———. "Tuvalu and Climate Change: Constructions of Environmental Displacement in Sydney Morning Herald." *Geografiska Annaler; Series B Human Geography* 87 (2005): 279–93.

———. "Wishful Sinking: Disappearing Islands, Climate Refugees and Cosmopolitan Experimentation." *Asia Pacific Viewpoint* 51, no. 1 (2010): 47–60.

Farbotko, Carol and Heather Lazrus. "The First Climate Refugees? Contesting Global Narratives of Climate Change in Tuvalu." *Global Environmental Change* 22, no. 2 (2012): 382–90.

Felli, Romain. "Managing Climate Insecurity by Ensuring Continuous Capital Accumulation: 'Climate Refugees' and 'Climate Migrants.'" *New Political Economy* 18, no. 1 (2012): 337–63.

Felli, Romain and Noel Castree. "Neoliberalising Adaptation to Climate Change: Foresight or Foreclosure?" *Environment and Planning A* 44, no. 1 (2012): 1–4.

Fields, Barbara Jeanne. "Slavery, Race, and Ideology in the United States of America." *New Left Review* 181, no. 1 (1990): 95–118.

Foucault, Michel. *Abnormal: Lectures at the Collège de France, 1974–1975*. New York: Picador, 2003.

———. "Confessions of the Flesh." In *Power/Knowledge: Selected Interviews and Other Writings, 1972–1977*, edited by Colin Gordon, 194–228. New York: Pantheon Books, 1980.

———. *The History of Sexuality: Volume 1: An Introduction*. New York: Vintage Books, 1978.

———. *"Society Must Be Defended": Lectures at the Collège de France, 1975–1976*. New York: Picador, 2003.

Fraser, Nancy. "Roepke Lecture in Economic Geography—from Exploitation to Expropriation: Historic Geographies of Racialized Capitalism." *Economic Geography* 94, no. 1 (2018): 1–17.

Frankenberg, Ruth. *White Women, Race Matters: The Social Construction of Whiteness*. Minneapolis: University of Minnesota Press, 1993.

Garner, Steve. *Racisms: An Introduction*. Thousand Oaks, CA: Sage, 2010.

Gemenne, François. "Climate-Induced Population Displacements in a 4 Degree Celsius+ World." *Philosophical Transactions of the Royal Society London: Mathematical, Physical and Engineering Sciences Series A* 369, no. 1934 (2011): 182–95.

———. "How They Became the Human Face of Climate Change. Research and Policy Interactions in the Birth of the 'Environmental Migration' Concept." In *Migration and Climate Change*, edited by Étienne Piguet, Antoine Pécoud and Paul de Guchteneire, 225–59. Cambridge: UNESCO Publishing and Cambridge University Press, 2011.

———. "Migration Doesn't Have to Be a Failure to Adapt: An Escape from Environmental Determinism." In *Climate Adaptation Futures*, edited by Jean Palutikof, Sarah Boulter, Andrew Ash, Mark Smith, Martin Parry, Marie Waschka and Daniela Guitart, 235–41. Oxford: Wiley-Blackwell, 2013.

———. "Why the Numbers Don't Add Up: A Review of Estimates and Predictions of People Displaced by Environmental Changes." *Global Environmental Change* 21, no. S1 (2011): S4–9.

Gemenne, François, Jon Barnett, Neil Adger and Geoffrey D. Dabelko. "Climate and Security: Evidence, Emerging Risks, and a New Agenda." *Climatic Change* 123, no. 1 (2014): 1–9.

Ghosh, Amitav. *The Great Derangement: Climate Change and the Unthinkable*. Chicago, IL: University of Chicago Press, 2016.

Gilbert, Emily. "The Militarization of Climate Change." *ACME* 11, no. 1 (2012): 1–14.

Gilmore, Ruthie. *Golden Gulag: Prisons, Surplus, Crisis, and Opposition in Globalizing California.* Berkeley: University of California Press, 2007.

Gilroy, Paul. *Darker than Blue: On the Moral Economies of Black Atlantic Culture*. Cambridge, MA: Harvard University Press, 2011.

———. *Postcolonial Melancholia*. New York: Columbia University Press, 2005.

———. *There Ain't No Black in the Union Jack*. London: Routledge, 1991.

Giroux, Henry. "Reading Hurricane Katrina: Race, Class, and the Biopolitics of Disposability." *College Literature* 33, no. 3 (2006): 172–96.

Giroux, Susan. "Sade's Revenge: Racial Neoliberalism and the Sovereignty of Negation." *Patterns of Prejudice* 44, no. 1 (2010): 1–26.

Giuliani, Gaia. *Monsters, Catastrophes and the Anthropocene: A Postcolonial Critique*. London: Routledge, 2020.

GMC. "Postcards from the Future." Last accessed 5 November 2021. http://www.postcardsfromthefuture.com/.

Goldberg, David Theo. *Are We All Post-Racial Yet?* Cambridge: Polity Press, 2015.

———. "Call and Response." *Patterns of Prejudice* 44, no. 1 (2010): 89–106.

———. “Racial Rule.” In *Relocating Postcolonialism: A Critical Reader*, edited by David Theo Goldberg and Atto Quayson, 82–102. Oxford: Blackwell Publishers, 2002.

———. *The Racial State*. Malden, MA: Blackwell, 2002.

———. *Racist Culture: Philosophy and the Politics of Meaning*. Oxford: Blackwell, 1993.

———. *The Threat of Race: Reflections on Racial Neoliberalism.* Oxford: Wiley-Blackwell, 2009.

Gordon, Jane and Lewis Gordon. *Of Divine Warning: Reading Disaster in a Modern Age*. Boulder, CO: Paradigm Publishers, 2009.

Government of the People’s Republic of Bangladesh. *Bangladesh Climate Change Strategy and Action Plan.* Dhaka: Ministry of Environment and Forests, 2009.

Grant, Madison. *The Passing of the Great Race*. New York: Charles Scribner’s Sons, 1916.

Gregg, Melissa and Gregory Seigworth. *The Affect Theory Reader*. Durham, NC, and London: Duke University Press, 2010.

Gregory, Derek. *The Colonial Present.* Oxford: Blackwell, 2004.

Grove, Jarius Victor. *Savage Ecology: War and Geopolitics at the End of the World.* Durham, NC, and London: Duke University Press, 2019.

Grusin, Richard. *Premediation: Affect and Mediation after 9/11*. London: Palgrave, 2010.

Gunaratnam, Yasmin and Nigel Clark. “Pre-Race, Post-Race: Climate Change and Planetary Humanism.” *Darkmatter Journal* 9, no. 1 (2012): n.p..

Gutman, Rachel. “The Three Most Chilling Conclusions from the Climate Report.” *The Atlantic*, 26 November 2018.

Haas, Peter M. “Introduction: Epistemic Communities and International Policy Coordination.” *International Organization* 46, no. 1 (1992): 1–35.

Hage, Ghassan. *Is Racism an Environmental Threat?* Cambridge: Polity Press, 2017.

Hall, Stuart. *The Fateful Triangle: Race, Ethnicity, Nation*. Cambridge, MA, and London: Harvard University Press, 2017.

———. “Race, Articulation and Societies Structured in Dominance.” In *Sociological Theories: Race and Colonialism,* 305–45. Paris: UNESCO, 1980.

Haraway, Donna. *Primate Visions: Gender, Race and Nature in the World of Modern Science*. New York and London: Routledge, 1989.

———. *Simians, Cyborgs and Women: A Reinvention of Nature*. London: Routledge, 1991.

———. “Situated Knowledges: The Science Question in Feminism and the Privilege of Partial Perspective.” *Feminist Studies* 14, no. 3 (1988): 575–99.

Harney, Stefano and Fred Moten. *The Undercommons: Fugitive Planning and Black Study*. Brooklyn, NY: Minor Compositions, 2013.

Hartmann, Betsy. “Rethinking Climate Refugees and Climate Conflict: Rhetoric, Reality and the Politics of Policy Discourse.” *Journal of International Development* 22, no. 2 (2010): 233–46.

Hau’ofa, Epeli. “Our Sea of Islands.” *The Contemporary Pacific* 6, no. 1 (1994): 148–61.

Hesse, Barnor. "Racialized Modernity: An Analytics of White Mythologies." *Ethnic and Racial Studies* 30, no. 4 (2007): 643–63.

Hobbes, Thomas. *Leviathan* (1651). London: Penguin Classics, 1968.

Hobson, John. *The Eurocentric Conception of World Politics: Western International Theory, 1760–2010*. Cambridge: Cambridge University Press, 2012.

Hook, Derek. "Affecting Whiteness: Racism as Technology of Affect." London: LSE Research Online, 2005.

———. *Foucault, Psychology, and the Analytics of Power*. Basingstoke: Palgrave Macmillan, 2007.

Howell, Alison and Melanie Richter-Monpetit. "Racism in Foucauldian Security Studies: Biopolitics, Liberal War, and the Whitewashing of Colonial and Racial Violence." *International Political Sociology* 13, no. 1 (2019): 2–19.

Hugo, Graeme. "Future Demographic Change and Its Interactions with Migration and Climate Change." *Global Environmental Change* 21 (2011): S21–S33.

Hulme, Mike. "Reducing the Future to Climate: A Story of Climate Determinism and Reductionism." *Osiris* 26, no. 1 (2011): 245–66.

Hunter, Lori M., Jessie K. Luna and Rachel M. Norton. "Environmental Dimensions of Migration." *Annual Review of Sociology* 41, no. 1 (2015): 377–97.

Huntington, Ellsworth. *Civilization and Climate*. Third edition. London: Oxford University Press, 1924.

Intergovernmental Panel on Climate Change. *Economic and Social Dimensions of Climate Change: Contribution of Working Group II to the Second Assessment Report of the IPCC*. Cambridge: Cambridge University Press, 1995.

———. *Fifth Assessment Report, Working Group II: Impacts, Adaptation, and Vulnerability*. Geneva: Intergovernmental Panel on Climate Change, 2014.

———. *Policymaker Summary of Working Group II (Potential Impacts of Climate Change)*. Bonn: Intergovernmental Panel on Climate Change, 1990.

International Organization for Migration. *The Atlas of Environmental Migration*. Abingdon: Routledge, 2017.

Ionesco, Dina, Daria Mokhnacheva and François Gemenne. *The Atlas of Environmental Migration*. London and New York: International Organization for Migration; Routledge, 2017.

Jakobeit, Cord and Chris Methmann. "'Climate Refugees' as Dawning Catastrophe? A Critique of the Dominant Quest for Numbers." In *Climate Change, Human Security, and Violent Conflict: Challenges for Societal Stability*, edited by Jürgen Scheffran, Michael Brzoska, Hans Günter Brauch, Peter Michael Link and Janpeter Schilling, 301–14. Berlin: Springer, 2012.

Jazeel, Tariq. "Spatializing Difference Beyond Cosmopolitanism: Rethinking Planetary Futures." *Theory, Culture & Society* 28, no. 5 (2011): 75–97.

Johnson, Elizabeth and Jesse Goldstein. "Biomimetic Futures: Life, Death, and the Enclosure of a More-than-Human Intellect." *Annals of the Association of American Geographers* 105, no. 2 (2015): 387–96.

Johnson, Leigh. "Catastrophe Bonds and Financial Risk: Securing Capital and Rule through Contingency." *Geoforum* 45 (March 2013): 30–40.

———. "Geographies of Securitized Catastrophe Risk and the Implications of Climate Change." *Economic Geography* 90, no. 2 (2014): 155–85.

Jones, Branwen Gruffydd. "'Good Governance' and 'State Failure': The Pseudo-Science of Statesmen in Our Times." In *Race and Racism in International Affairs*, edited by Alexander Anievas, Nivi Manachanda and Robbie Shilliam, 62–80. London: Routledge 2015.

———. "Race in the Ontology of International Order." *Political Studies* 56, no. 4 (2008): 907–27.

Kaplan, Robert. "Waterworld." *The Atlantic*, 15 February 2008.

Karera, Axelle. "Blackness and the Pitfalls of Anthropocene Ethics." *Critical Philosophy of Race* 7, no. 1 (2019): 32–56.

Kelley, Colin P., Shahrzad Mohtadi, Mark A. Cane, Richard Seager and Yochanan Kushnir. "Climate Change in the Fertile Crescent and Implications of the Recent Syrian Drought." *Proceedings of the National Academy of Sciences of the United States of America* 112, no. 11 (2015): 3241–46.

Khanna, Parag. "Migration Will Soon Be the Biggest Challenge of Our Time." *Financial Times*, 3 October 2021.

Klein, Naomi. *This Changes Everything: Capitalism vs. The Climate.* London: Simon and Schuster, 2014.

———. "Let Them Drown: The Violence of Othering in a Warming World." *London Review of Books* 38, no. 11 (2 June 2016), 11–13.

Kniveton, Dominic, Kerstin Shmidt-Verkek, Christopher Smith and Richard Black. *Climate Change and Migration: Improving Methodologies to Estimate Flows*. Geneva: International Organization of Migration, 2008.

Kobayashi, Audrey. "The Construction of Geographical Knowledge—Racialization, Spatialization." In *Handbook of Cultural Geography*, edited by Kay Anderson, Mona Domosh, Steve Pile and Nigel Thrift, 544–56. Thousand Oaks, CA: Sage, 2003.

Kobayashi, Audrey and Linda Peake. "Racism Out of Place: Thoughts on Whiteness and Anti-Racist Geography in the New Millennium." *Annals of the American Association of Geographers* 90, no. 2 (June 2000): 392–403.

———. "Unnatural Discourse: 'Race' and Gender in Geography." *Gender, Place and Culture* 1, no. 2 (1994): 225–43.

Kolmannskog, Vikram Odedra. *Future Floods of Refugees*. Oslo: Norwegian Refugee Council, 2008.

Kothari, Uma. "An Agenda for Thinking about 'Race' in Development." *Progress in Development Studies* 6, no. 1 (2006): 9–23.

Kundnani, Arun. "The Racial Constitution of Neoliberalism." *Race and Class* 63, no. 1 (April 2021): 51–69.

———."What Is Racial Capitalism?" Public Lecture, Havens Wright Center for Social Justice, University of Wisconsin-Madison, 15 October 2020.

Laczko, Frank and Christine Aghazarm, eds. *Migration, Environment and Climate Change: Assessing the Evidence.* Geneva: International Organization for Migration, 2009.

Lake, Marilyn and Henry Reynolds. *Drawing the Global Colour Line: White Men's Countries and the International Challenge of Racial Equality*. Cambridge: Cambridge University Press, 2008.
Latour, Bruno. *The Politics of Nature: How to Bring the Sciences into Democracy*. Cambridge, MA: Harvard University Press, 2004.
———. *We Have Never Been Modern*. New York: Harvester/Wheatsheaf, 1993.
Lentin, Alana. "Beyond Denial: 'Not Racism' as Racist Violence." *Continuum* 32, no. 4 (2018): 400–14.
———. *Why Race Still Matters*. London: Polity Press, 2020.
Lévi-Strauss, Claude. *Triste Tropical*. New York: Atheneum, 1973.
Livingstone, David. "Environmental Determinism." In *The Sage Handbook of Geographical Knowledge*, edited by John Agnew and David Livingstone, 368–80. Thousand Oaks, CA: Sage, 2011.
———. *The Geographical Tradition: Episodes in the History of a Contested Enterprise*. Oxford: Blackwell, 1992.
Lobo-Guerrero, Luis. *Insuring Security: Biopolitics, Security and Risk.* London: Routledge, 2011.
Losurdo, Domenico. *Liberalism: A Counterhistory*. London: Verso, 2014.
Lowe, Lisa. *The Intimacies of Four Continents*. Durham, NC, and London: Duke University Press, 2015.
Luke, Timothy W. "The Climate Change Imaginary." *Current Sociology* 63, no. 2 (2015): 280–96.
———. "Tracing Race, Ethnicity, and Civilization in the Anthropocene." *Environment and Planning D: Society and Space* 38, no. 1 (2020): 129–46.
Lustgarten, Abrahm. "The Great Climate Migration Has Begun." *The New York Times Magazine*, 23 July 2020.
———. "How Climate Change Will Reshape America." *The New York Times Magazine*, 15 September 2020.
———. "How Russia Wins the Climate Crisis." *The New York Times Magazine*, 16 December 2020.
Mackey, Eva. *The House of Difference: Cultural Politics and National Identity in Canada*. Toronto: University of Toronto Press, 2002.
Mahony, Martin. "Picturing the Future-Conditional: Montage and the Global Geographies of Climate Change." *Geo: Geography and Environment* 3, e0009, no. 2 (2016): 1–18.
Maldonado, Julie and Kristina Peterson. "A Community-Based Model for Resettlement." In *The Handbook on Environmental Migration and Displacement*, edited by Robert McLeman and François Gemenne, 289–99. Abingdon: Routledge, 2018.
Manley, Gordon. "Some Recent Contributions to the Study of Climate Change." *Quarterly Journal of the Royal Meteorological Society* 70, no. 305 (July 1944): 197–219.
Mann, Geoff and Joel Wainwright. *Climate Leviathan: A Political Theory of Our Planetary Future*. London: Verso, 2018.

Marino, Elizabeth. "The Long History of Environmental Migration: Assessing Vulnerability Construction and Obstacles to Successful Relocation in Shishmaref, Alaska." *Global Environmental Change* 33, no. 2 (May 2012): 374–81.

Marriot, David. "Inventions of Existence: Sylvia Wynter, Frantz Fanon, Sociogeny and the 'Damned.'" *CR: The New Centennial Review* 11, no. 3 (Winter 2011): 45–89.

Massumi, Brian. "National Enterprise Emergency: Steps towards an Ecology of Powers." *Theory, Culture and Society* 26, no. 6 (2009): 153–85.

———. *Ontopower: War, Powers, the State of Perception.* Durham, NC, and London: Duke University Press, 2015.

———. *Parables for the Virtual: Movement, Affect, Sensation.* Durham, NC, and London: Duke University Press, 2002.

———. "Potential Politics and the Primacy of Preemption." *Theory and Event* 10, no. 2 (2007): n.p.

Mayer, Benoit. "The International Legal Challenges of Climate-Induced Migration: Proposal for an International Legal Framework." *Colorado Journal of International Environmental Law and Policy* 22 (2011): 357–416.

Mbembe, Achille. "Necropolitics." *Public Culture* 15, no. 1 (2003): 11–40.

McAdam, Jane. *Climate Change, Forced Migration, and International Law*. Oxford: Oxford University Press, 2012.

———. "'Disappearing States,' Statelessness and the Boundaries of International Law." In *Climate Change and Displacement: Multidisciplinary Perspectives*, edited by Jane McAdam, 105–29. Oxford: Hart Publishing, 2010.

———. "From the Nansen Initiative to the Platform on Disaster Displacement: Shaping International Approaches to Climate Change, Disasters and Displacement." *University of New South Wales Law Journal*, no. 39 (2016): 1518–46.

———. "Refusing 'Refuge' in the Pacific." In *Migration and Climate Change*, edited by Étienne Piguet, Antoine Pécoud and Paul Gucheteneire, 102–37. Cambridge: Cambridge University Press, 2011.

McAdam, Jane and Ben Saul. "Displacement with Dignity: International Law and Policy Responses to Climate Change Migration and Security in Bangladesh." *German Yearbook of International Law* 53 (2010): 233–87.

———. "An Insecure Climate for Human Security? Climate-Induced Displacement and International Law." In *Human Security and Non-Citizens: Law, Policy and International Affairs*, edited by Alice Edwards, 357–403. Cambridge: Cambridge University Press, 2009.

McCarthy, Jesse. "On Afropessimism." *Los Angeles Review of Books*, 20 July 2020.

McClintock, Anne. *Imperial Leather: Race, Gender and Sexuality in the Colonial Contest*. New York: Routledge, 1995.

McDonald, Matt. "Discourses of Climate Security." *Political Geography* 33 (March 2013): 42–51.

McGranahan, Gordon. "The Rising Tide: Assessing the Risks of Climate Change and Human Settlements in Low Elevation Coastal Zones." *Environment and Urbanization* 19, no. 1 (2007): 17–37.

McKee, Yates. "On Climate Refugees: Biopolitics, Aesthetics, and Critical Climate Change." *Qui Parle* 19, no. 2 (Spring/Summer 2011): 309–325.

McKittrick, Katherine. "Plantation Futures." *Small Axe* 17, no. 3 (2013): 1–15.

McLaren, Angus. *Our Own Master Race: Eugenics in Canada, 1885–1945*. Toronto: University of Toronto Press, 2015.

McLeman, Robert. *Climate and Human Migration: Past Experiences, Future Challenges*. Cambridge: Cambridge University Press, 2014.

———. "Developments in Modelling Climate Change-Related Migration." *Climatic Change* 117, no. 3 (2013): 599–611.

McLeman, Robert and Barry Smit. "Migration as an Adaptation to Climate Change." *Climatic Change* 76, no. 1 (2006): 31–53.

McLeman, Robert and François Gemenne. "Environmental Migration Research: Evolution and Current State of the Science." In *Routledge Handbook of Environmental Migration and Displacement*, edited by Robert McLeman and François Gemenne, 3–16. Abingdon: Routledge, 2018.

McLeman, Robert, Mohammad Moniruzzaman and Nasima Akter. "Environmental Influences on Skilled Worker Migration from Bangladesh to Canada." *The Canadian Geographer* 62, no. 3 (2018): 352–71.

McNamara, Karen. "Cross-Border Migration with Dignity in Kiribati." *Forced Migration Review* 49 (May 2015): 62.

McNamara, Karen and Chris Gibson. "'We Do Not Want to Leave Our Land': Pacific Ambassadors at the United Nations Resist the Category of 'Climate Refugees.'" *Geoforum* 40 (2009): 475–83.

McWhorter, Ladelle. *Racism and Sexual Oppression in America: A Genealogy.* Bloomington: Indiana University Press, 2009.

Methmann, Chris and Angela Oels. "From 'Fearing' to 'Empowering'Climate Refugees: Governing Climate-Induced Migration in the Name of Resilience." *Security Dialogue* 46, no. 1 (February 2015): 51–68.

Meyer, William B. and Dylan M. T. Guss. *Neo-Environmental Determinism: Geographical Critiques.* London: Palgrave Macmillan, 2017.

Mikkelsen, Jon, ed. *Kant and the Concept of Race: Late Eighteenth-Century Writings*. Albany: State University of New York Press, 2013.

Miller, Todd, Nick Buxton and Mark Akkerman. *Global Climate Wall: How the World's Wealthiest Nations Prioritize Borders over Climate Action.* Amsterdam: Transnational Institute, 2021.

Millman, Oliver. "Climate Denial Is Waning on the Right. What's Replacing It Might Be Just as Scary." *The Guardian*, 21 November 2021.

Mills, Charles. "Global White Ignorance." In *Routledge Handbook of Ignorance Studies*, edited by Matthias Gross and Linsey McGoey, 217–27. London: Routledge, 2015.

Mirzoeff, Nicholas. "It's Not the Anthropocene, It's the White Supremacy Scene; or the Geological Color Line." In *After Extinction*, edited by Richard Grusin, 123–50. Minneapolis: University of Minnesota Press, 2018.

Moore, Donald, Jake Kosek and Anand Pandian. "The Cultural Politics of Race and Nature: Terrains of Power and Practice." In *Race, Nature and the Politics of Difference*, edited by Donald Moore, Jake Kosek and Anand Pandian, 1–70. Durham, NC, and London: Duke University Press, 2003.

Morrison, Toni. *Playing in the Dark: Whiteness and the Literary Imagination.* London: Picador, 1993.

Morrissey, James. *Environmental Change and Forced Migration: A State of the Art Review.* Discussion paper. Oxford: Refugee Studies Centre, Oxford University, 2009.

———. "Rethinking the 'Debate on Environmental Refugees': From 'Maximalists and Minimalists' to 'Proponents and Critics.'" *Journal of Political Ecology* 19 (2012): 36–49

Moya, Paula M. L. "The Search for Decolonial Love: An Interview with Junot Díaz." *Boston Review*, 26 June 2012.

Muret, Maurice. *The Twilight of the White Races.* New York: Charles Scribner's Sons, 1926.

Mwape, Darlington and Volker Türk. *Report by the Co-Chairs: Gaps in the International Protection Framework and in Its Implementation.* Geneva: United Nations High Commissioner on Refugees, 2010.

Myers, Norman. "Environmental Refugees: An Emergent Security Issue." Conference paper. Prague: World Economic Forum, 2005.

———. "Environmental Refugees in a Globally Warmed World." *BioScience* 43, no. 11 (1993): 752–61.

Nail, Thomas. *The Figure of the Migrant.* Stanford, CA: Stanford University Press, 2016.

Nash, Michael, director. *Climate Refugees: The Human Face of Climate Change.* LA Think Tank, 2010, 89 mins.

Nash, Sarah. *Negotiating Migration in the Context of Climate Change: The Coming of Age of an Area of Policymaking.* Bristol: Bristol University Press, 2019.

Nelson, Sara. "Beyond the Limits to Growth: Ecology and the Neoliberal Counterrevolution." *Antipode* 47, no. 2 (2015): 461–80.

———. "Resilience and the Neoliberal Counter-Revolution: From Ecologies of Control to the Production of the Common." *Resilience* 2, no. 1 (January 2014): 1–17.

Nicholls, Robert. "Coastal Flooding and Wetland Loss in the 21st Century: Changes under the SRES Climate and Socio-Economic Scenarios." *Global Environmental Change* 14, no. 1 (2004): 69–86.

Nicholson, Callum. "Climate Change and the Politics of Causal Reasoning: The Case of Climate Change and Migration." *The Geographical Journal* 180, no. 2 (2014): 151–60.

Noxolo, Patricia, Parvati Ragurham and Clare Madge. "Unsettling Responsibility: Postcolonial Interventions." *Transactions of the Institute of British Geographers* 37, no. 3 (2012): 418–29.

O'Hagen, Ellie Mae. "Mass Migration Is No 'Crisis': It's the New Normal as the Climate Changes." *The Guardian*, 18 August 2015.

Omi, Michael and Howard Winant. *Racial Formation in the United States: From the 1960s to the 1990s.* Third edition. New York: Routledge, 2015.

Pailey, Robtel Neajai. "De-Centring the 'White Gaze' of Development." *Development and Change* 51, no. 3 (May 2020): 729–45.

Panayi, Panikos. *An Immigration History of Britain: A Multicultural History of Racism since 1800*. New York and London: Routledge, 2010.

Paprocki, Kasia. "All that is solid melts into the bay: Anticipatory ruination and climate change adaptation." *Antipode* 51, no. 1 (2019): 295–315.

Parker, Laura. "143 Million People May Soon Become Climate Migrants." *National Geographic*, 19 March 2018.

Pearson, Charles. *National Life and Character: A Forecast*. London: Macmillan and Co., 1893.

Peck, Raoul, director, *I Am Not Your Negro.* Velvet Film & Magnolia Pictures, 2016, 95 mins.

Peet, Richard. "The Social Origins of Environmental Determinism." *Annals of the Association of American Geographers* 75, no. 3 (1985): 309–33.

Piguet, Etienne. "From 'Primitive Migration' to 'Climate Refugees': The Curious Fate of the Natural Environment in Migration Studies." *Annals of the Association of American Geographers* 103, no. 1 (2014): 148–62.

Pollard, Jane, Jonathan Oldfield, Sam Randalls and John Thornes. "Firm Finances, Weather Derivatives, and Geography." *Geoforum* 39, no. 2 (March 2008): 612–24.

Pratt, Mary Louise. *Imperial Eyes: Travel Writing and Transculturation*. London: Routledge, 1992.

Pred, Allen. *Even in Sweden: Racisms, Racialized Spaces, and the Popular Geographic Imagination*. Oakland: University of California Press, 2000.

Prevention Web. "United Nations Discussion Panel: Insure Me: Climate Change, Human Migration, and Risk, September 24, 2009." Last accessed 16 November 2021. https://www.preventionweb.net/events/view/11131?id=11131.

Protevi, John. *Political Affect: Connecting the Social and the Somatic*. Minneapolis: University of Minnesota Press, 2008.

Pulido, Laura. "Flint, Environmental Racism, and Racial Capitalism." *Capitalism Nature Socialism* 27, no. 3 (2016): 1–16.

————. "Geographies of Race and Ethnicity II: Environmental Racism, Racial Capitalism and State-Sanctioned Violence." *Progress in Human Geography* 41, no. 4 (2017): 524–33.

Radcliffe, Sarah, Elizabeth Watson, Ian Simmons, Felipe Fernández-Arnesto and Andrew Sluyter. "Environmentalist Thinking and/in Geography." *Progress in Human Geography* 34, no. 1 (2010): 98–116.

Randalls, Sam. "Weather Profits: Weather Derivatives and the Commercialization of Meteorology." *Social Studies of Science* 40, no. 5 (2010): 705–30.

Razack, Sherene. *Casting Out: The Eviction of Muslims from Western Law and Politics*. Toronto: University of Toronto Press, 2008.

Reuveny, Rafael. "Climate Change-Induced Migration and Violent Conflict." *Political Geography* 26, no. 6 (2007): 656–73.

Rickards, Lauren, John Wiseman and Taegen Edwards. "The Problem of Fit: Scenario Planning and Climate Change Adaptation in the Public Sector." *Environment and Planning C: Government and Policy* 32, no. 4 (2014): 641–62.

Rigaud, Kumari, Alex de Sherbinin, Bryan Jones, Jonas Bergmann, Viviane Clement, Kayly Ober, Jacob Schewe, Susana Adamo, Brent McCusker, Silke Heuser and Amelia Midgley. *Groundswell: Preparing for Internal Climate Migration.* Washington, DC: The World Bank, 2018.

Rintoul, Douglas. *The Edge*. Transport Theatre, 2015.

Roberts, David and Minelle Mahtani. "Neoliberalizing Race, Racing Neoliberalism: Placing 'Race' in Neoliberal Discourses." *Antipode* 42, no. 2 (2010): 248–57.

Robinson, Cedric. *Black Marxism: The Making of the Black Radical Tradition.* Chapel Hill and London: University of North Carolina Press, 2000.

Roediger, David. *The Wages of Whiteness: Race and the Making of the American Working Class*. London and New York: Verso, 1991.

Ross, Tim. "Boris Johnson Likens Climate Change to 'the Decline and Fall of the Roman Empire.'" *The New Statesman*, 29 October 2021.

Said, Edward. *Culture and Imperialism*. New York: Vintage Books, 1994.

———. *Orientalism*. New York: Vintage Books, 1979.

Saldanha, Arun. "A Date with Destiny: Racial Capitalism and the Beginnings of the Anthropocene." *Environment and Planning D: Society and Space* 38, no. 1 (2020): 12–34.

———. *Psychedelic White: Goa Trace and the Viscosity of Race*. Minneapolis: University of Minnesota Press, 2007.

Saunders, Patricia. "Environmental Refugees: The Origins of a Construct." In *Political Ecology: Science, Myth and Power*, edited by Philip Stott and Sian Sullivan, 218–46. Oxford: Oxford University Press, 2000.

Schmitt, Carl. *The Nomos of the Earth* (1950). Translated by G. L. Ulmen. New York: Telos Press, 2003.

Schwartz, Peter and Doug Randall. *An Abrupt Climate Change Scenario and Its Implications for United States National Security*. Emeryville: Global Business Network, 2003.

Schwarz, Bill. *Memories of Empire, Volume 1, the White Man's World.* Oxford: Oxford University Press, 2011.

Scott, Jared, director. *The Age of Consequences*. Moon Atlas and PF Pictures, 2016, 80 mins.

Sealey-Huggins, Leon. "Between Apocalypse and Survival: The Violence of Climate Breakdown in, and for, the Caribbean." Conference Keynote Lecture. *Political Ecology of the Far-Right*, 15–17 November 2019. Lund: Lund University.

Selby, Jan, Omar Dahi, Christiane Fröhlich and Mike Hulme. "Climate Change and the Syrian Civil War Revisited." *Political Geography* 60 (2017): 232–44.

Seth, Sayjay, ed. *Postcolonial Theory and International Relations*. Abingdon: Routledge, 2013.

Sexton, Jared. "On Black Negativity, or the Affirmation of Nothing." Interview by Daniel Barber, *Society and Space*, 18 September 2017.

Shapiro, Michael. *Violent Cartographies: Mapping Cultures of War*. Minneapolis: University of Minnesota Press, 1997.

Sharpe, Christina. *In the Wake: On Blackness and Being*. Durham, NC, and London: Duke University Press, 2016.

Shields, Rob. *The Virtual*. London and New York: Routledge, 2003.
Shilliam, Robbie. *Race and the Undeserving Poor: From Abolition to Brexit*. London: Agenda Publishing, 2017.
Siddiqui, Tasneem, Mohammad Towheedul Islam and Zohra Akhter. *National Strategy on the Management of Disaster and Climate Induced Internal Displacement*. Dhaka, Bangladesh: Ministry of Disaster Management and Relief, 2015.
Sluyter, Andrew. "Neo-Environmental Determinism, Intellectual Damage Control, and Nature/Society Science." *Antipode* 35, no. 4 (2003): 813–17.
Spengler, Oswald. *The Decline of the West*. London: George Allen and Unwin, 1932.
Stein, Michael Isaac. "How to Save a Town from Rising Waters." *Wired Magazine*, 25 January 2018.
Stengers, Isabelle. *In Catastrophic Times: Resisting the Coming Barbarism*. London: Open Humanities Press, 2015.
Stern, Sir Nicholas. *Stern Review on the Economics of Climate Change*. London: HM Treasury, 2006.
Stoddard, Lothrop. *The Rising Tide of Color against White World-Supremacy*. New York: Charles Scribner's Sons, 1925.
Stojanov, Robert, Barbora Duží, Ilan Kelman, Daniel Němec and David Procházka. "Local Perceptions of Climate Change Impacts and Migration Patterns in Malé, Maldives." *The Geographical Journal* 183, no. 4 (2016): 370–85.
Stoler, Ann Laura. *Duress: Imperial Durabilities in Our Times*. Durham, NC, and London: Duke University Press, 2016.
———. *Race and the Education of Desire: Foucault's History of Sexuality and the Colonial Order of Things*. Durham, NC, and London: Duke University Press, 1995.
Swyngedouw, Erik. "Antimonies of the Postpolitical City: In Search of a Democratic Politics of Environmental Production." *International Journal of Urban and Regional Research* 33, no. 3 (2009): 601–20.
Telford, Andrew. "A Climate Terrorism Assemblage? Exploring the Politics of Climate Change-Terrorism-Radicalisation Relations." *Political Geography* 79 (May 2020): 102–50.
Tesfahuney, Mekkonen. "Mobility, Racism and Geopolitics." *Political Geography* 17, no. 5 (1998): 499–515.
Theweleit, Klaus. *Male Fantasies: Volume 1: Women, Floods, Bodies, Histories*. Minneapolis: University of Minnesota Press, 1987.
Thobani, Sunera. *Exalted Subjects: Studies in the Making of Race and Nation in Canada*. Toronto and Buffalo, NY: University of Toronto Press, 2007.
Todd, Zoe. "Indigenizing the Anthropocene." In *Art in the Anthropocene: Encounters among Aesthetics, Politics, Environments and Epistemologies*, edited by Heather Davis and Etienne Turpin, 241–54. London: Open Humanities Press, 2015.
Tooze, Adam. "The COP26 Message? We Are Trusting Big Business, Not States, to Fix the Climate Crisis." *The Guardian*, 16 November 2021.
Tuana, Nancy. "Climate Apartheid: The Forgetting of Race in the Anthropocene." *Critical Philosophy of Race* 7, no. 1 (2019): 1–31.
Tuck, Eve and K. Wayne Yang. "Decolonization Is Not a Metaphor." *Decolonization: Indigeneity, Education and Society* 1, no. 1 (2012): 1–40.

U.K. Foresight. *Migration and Global Environmental Change: Final Project Report.* London: The Government Office for Science, 2011.

United Nations Framework Convention on Climate Change. *The Cancun Adaptation Agreements*. FCCC/CP/2010/7/Add.1. Decision 1/CP.16. Cancun: UNFCCC, 2010.

United Nations High Commissioner for Refugees. *Convention and Protocol Relating to the Status of Refugees*. Geneva: United Nations High Commissioner for Refugees, 1951.

United Nations Security Council. *Letter Dated 5 April from the Permanent Representative of the United Kingdom of Great Britain and Northern Ireland to the United Nations Addressed to the President of the Security Council*, S/2007/186. New York: United Nations Security Council, 2007.

United Nations University Migration Network. "CCEMA." Last accessed 17 November 2021, https://migration.unu.edu/research/forced-migration/climate-change-environment-and-migration-alliance-ccema.html#outline.

Vitalis, Robert. *America's Kingdom: Mythmaking on the Saudi Oil Frontier*. Stanford, CA: Stanford University Press, 2008.

———. *White World Order, Black Power Politics: The Birth of American International Relations*. Ithaca, NY: Cornell University Press, 2015.

Walker, Barrington, ed. *The History of Immigration and Racism in Canada: Essential Readings*. Toronto: Canadian Scholars Press, 2008.

Walker, Jeremy and Melinda Cooper. "Genealogies of Resilience: From Systems Ecology to the Political Economy of Crisis Adaptation." *Security Dialogue* 42, no. 2 (April 2011): 143–60.

Wanjiku-Kelbert, Alexandra and Joshua Virasami. "Darkening the White Heart of the Climate Movement." *The New Internationalist*, 1 December 2015.

Ward, Deborah. *The White Welfare State: The Racialization of US Welfare Policy*. Ann Arbor: University of Michigan Press, 2005.

Warner, Koko. "Global Environmental Change and Migration: Governance Challenges." *Global Environmental Change* 20, no. 3 (2010): 402–13.

———. "Human Migration and Displacement in the Context of Adaptation to Climate Change: The Cancun Adaptation Framework and Potential for Future Action." *Environment and Planning C: Government and Policy* 30, no. 6 (2012): 1061–77.

Warner, Koko, Charles Ehrhart, Alex de Sherbinin, Susana Adamo and Tricia Chai-Onn. *In Search of Shelter: Mapping the Effects of Climate Change on Human Migration and Displacement.* Bonn: United Nations University, 2009.

Warren, Calvin L. *Ontological Terror: Blackness, Nihilism, and Emancipation.* Durham, NC: Duke University Press, 2018.

Watts, Michael. "On Confluences and Divergences." *Dialogues in Human Geography* 1, no. 1 (2011): 84–89.

Weheliye, Alexander. *Habeas Viscus: Racializing Assemblages, Biopolitics, and Black Feminist Theories of the Human*. Durham, NC, and London: Duke University Press, 2016.

Wekker, Gloria. "Afropessimism." *European Journal of Women's Studies* 28, no. 1 (2020): 86–97.

Wennersten, John and Denise Robbins. *Rising Tides: Climate Refugees in the Twenty-First Century*. Bloomington: Indiana University Press, 2017.

Werz, Michael and Laura Conley. *Climate Change, Migration and Conflict: Addressing Complex Crisis Scenarios in the 21st Century*. Washington, DC: Center for American Progress, 2012.

White, Gregory. *Climate Change and Migration: Security and Borders in a Warming World*. Oxford: Oxford University Press, 2011.

White, Sarah. “Thinking Race, Thinking Development.” *Third World Quarterly* 23, no. 3 (2002): 407–19.

Wilderson, Frank, III. *Afropessimism*. New York: Liveright, 2020.

———. *Red, White, and Black: Cinema and the Structure of US Antagonism*. Durham, NC, and London: Duke University Press, 2010.

Wiegman, Robyn. “Whiteness Studies and the Paradox of Particularity.” *Boundary 2* 26, no. 3 (Autumn 1999): 115–50.

Williams, Angela. “Turning the Tide: Recognizing Climate Change Refugees in International Law.” *Law and Policy* 30, no. 4 (2008): 502–29.

Wilson, Kalpana. “‘Race,’ Gender and Neoliberalism: Changing Visual Representations in Development.” *Third World Quarterly* 32, no. 2 (2011): 315–31.

Wretched of the Earth. “Open Letter from the Wretched of the Earth to the Organisers of the People’s Climate March of Justice and Jobs.” *Dissidents*, 16 December 2015.

Wynter, Sylvia. “Towards the Sociogenic Principle: Fanon, Identity, the Puzzle of Conscious Experience, and What It Is Like to Be ‘Black.’” *National Identities and Sociopolitical Changes in Latin America* (2001): 30–66.

———. “Unsettling the Coloniality of Being/Power/Truth/Freedom: Towards the Human, after Man, Its Overrepresentation—an Argument.” *CR: The New Centennial Review* 3, no. 3 (2003): 257–337.

Younis, Musab. “‘United by Blood’: Race and Transnationalism During the Belle Époque.” *Nations and Nationalism* 23, no. 3 (2017): 484–504.

Yusoff, Kathryn. *A Billion Black Anthropocenes or None*. Minneapolis: University of Minnesota Press, 2018.

Yusoff, Kathryn and Jennifer Gabrys. “Climate Change and the Imagination.” *Wiley Interdisciplinary Reviews: Climate Change* 2 (July/August 2011): 516–34.

# Index

# About the Author

**Andrew Baldwin** is associate professor of human geography at Durham University. He is the co-editor of *Life Adrift: Climate Change, Migration, Critique* and *Climate Change, Migration and Human Rights: Law and Policy Perspectives*.